T0212398

# THE FRONTIERS COLLECTION

The books in this collection are devoted to challenging and open problems at the forefront of modern science and scholarship, including related philosophical debates. In contrast to typical research monographs, however, they strive to present their topics in a manner accessible also to scientifically literate non-specialists wishing to gain insight into the deeper implications and fascinating questions involved. Taken as a whole, the series reflects the need for a fundamental and interdisciplinary approach to modern science and research. Furthermore, it is intended to encourage active academics in all fields to ponder over important and perhaps controversial issues beyond their own speciality. Extending from quantum physics and relativity to entropy, consciousness, language and complex systems—the Frontiers Collection will inspire readers to push back the frontiers of their own knowledge.

More information about this series at http://www.springer.com/series/5342

Giovanni Battimelli · Giovanni Ciccotti ·
Pietro Greco

# Computer Meets Theoretical Physics

The New Frontier of Molecular Simulation

 Springer

Giovanni Battimelli
Department of Physics
Sapienza University of Rome
Rome, Italy

Pietro Greco
Città della Scienza
Naples, Italy

Giovanni Ciccotti
Department of Physics
Sapienza University of Rome
Rome, Italy

IAC "Mauro Picone" CNR
Rome, Italy

School of Physics
University College of Dublin
Dublin, Ireland

*Translated by*
Giuliana Giobbi
Segreteria Scientifica INAF-OAR
Osservatorio Astronomico di Roma
Monte Porzio Catone, Italy

G. Battimelli, G. Ciccotti, P. Greco: Il computer incontra la fisica teorica. La nuova frontiera della simulazione molecolare, Carocci editore, Roma (2020)

ISSN 1612-3018          ISSN 2197-6619   (electronic)
THE FRONTIERS COLLECTION
ISBN 978-3-030-39401-1         ISBN 978-3-030-39399-1   (eBook)
https://doi.org/10.1007/978-3-030-39399-1

This Springer imprint is published by the registered company Springer Nature Switzerland AG
The registered company address is: Gewerbestrasse 11, 6330 Cham, Switzerland

# Preface

The idea of this book first took shape in the head of one of the authors (GC) following a conversation with his late friend Pierre Turq, a chemist from Paris who was among the first scientists in France to foresee the potentialities of computer simulation for the development of his own discipline. We wondered what, if any, mysterious reasons lay behind the fact that the epistemological upheaval brought by molecular simulation techniques into physical chemistry had never found a counterpart in some form of public exposition aimed at telling its history and explaining its relevance. Pierre felt that this circumstance was rooted in a sort of reluctance, by the chemistry community, to critically consider, and give a public account of, its own history. In so doing, chemistry would thereby admit, in the epistemological hierarchy of scientific disciplines, its subordinate position to physics, whose practitioners on the contrary have always been rather keen to give a public image of their science as the one best able to handle those problems that most deserve critical attention and historiographical scrutiny.

This seemed to be a puzzling contradiction, given that, even at the time of this conversation, simulation had long since asserted itself as a pervasive approach in all fields of theoretical physics, in spite of the fact that its roots, and its first founding fathers, were rather to be found in physical chemistry. The inconsequence of the circumstance was manifest: the creation of a growing set of ideas, techniques, and potentialities that had come to pervade the whole field of fundamental research, redefining the canonical boundaries between contiguous disciplines and radically altering their cognitive procedures, had been substantially ignored as a subject worthy of critical inquiry and public exposition because its origins were not deemed noble enough. An original sin condemning it to oblivion.

This seems all the more paradoxical because in the meantime the novelties brought about by the intensive use of computers have not been confined to the emergence of a new way of doing theoretical physics, and to its extension to research fields traditionally regarded as being outside the dominion of the "hard" sciences (such as biology or pharmaceutical chemistry), but have also produced a significant impact on mathematics, giving birth to new sectors of applied mathematics concerned with numerical analysis, mathematical statistics, and construction of the increasingly sophisticated algorithms needed for intensive computation.

Only in recent times have a few research projects on the history and assessment of simulation seen the light, along with a few works of historiographical analysis; they have, however, been confined to the academic environment and have in no way contributed to modifying the substantial ignorance on the matter in the wider scientific culture. Be this as it may, the final result is that an outstanding innovation in the practice of fundamental research, one that has substantially reshuffled the traditional disciplinary hierarchies, has grown up and established itself without a collective perception of that novelty in the image of science.

This book is an attempt to obviate this incongruence. We thought that the optimal way to proceed would be to present the reader with the historical development of this transformation, by keeping technical matters down to the indispensable minimum, and ending our narration at the moment when it can be considered that the potentialities emerging with the birth of molecular simulation, around the mid-50s, were fully expressed, and the development of the sector had given rise to a clearly defined discipline with its own identity; our reconstruction therefore stops at the end of the 80s, when this first phase of the history of molecular simulation reached its completion.

Rome, Italy                                                                                     Giovanni Battimelli
Rome, Italy                                                                                      Giovanni Ciccotti
Naples, Italy                                                                                         Pietro Greco

# Acknowledgements

Right through this work, we have been helped in various ways by so many friends and colleagues that we have almost certainly overlooked some names; we apologize in advance for any omission. We thank in particular Alin Elena for his careful reading of the text; Sara Bonella, Daniel Borgis, Ignacio Pagonabarraga, Benjamin Rotenberg, and Rodolphe Vuilleumier for their continuous support of the initiative; Mauro Ferrario for being systematically available to help; and Carlo Pierleoni for helping to orient us in the intricacies of the physics questions related to quantum Monte Carlo. Among the protagonists of our history, we have received useful informations and suggestions from Charles Bennett, Jean-Pierre Hansen, Ray Kapral, Mike Klein, Michel Mareschal, Ian McDonald, Thanos Panagiotopoulos, Daniel Schiff, and Glenn Torrie. For the reconstruction of the origins of our history, we have profited invaluably from the long interaction with Berni Alder. We are particularly indebted to Hans Christian Andersen, Kurt Binder, David Ceperley, Daan Frenkel, Mal Kalos, Dominique Levesque, and Mary Ann Mansigh; beside sending precious comments, they have repeatedly intervened directly in our original version, greatly improving the readability of the text and correcting slips and inaccuracies; Daan, in particular, deserves our gratitude for his systematic presence and support all along our adventure. It goes without saying that we take full responsibility for any remaining faults and omissions still present in the text. Part of this work has profited from the generous support of the Stefanovich Institute on the Formation of Knowledge (SIFK), Chicago, thanks to the wisdom of its Director, Professor Shadi Bartsch-Zimmer, and, more recently from that of the Neubauer Collegium, Chicago, for its Molecular Dynamics project. Last but not least, we wish to express our gratitude to Gianluca Mori from Carocci Editore, and to Chris Caron and Lisa Scalone, from Springer Verlag, for the encouragement and support given to the book proposal, to Giuliana Giobbi for the hard work put into the book's translation, and to Stephen Lyle for his careful revision.

A special word of gratitude to our wives who stoically endured the disorder that the production of this work induced into our family lives. More than that, GC is indebted to Nicoletta Bosisio for many insightful fights.

# Contents

# Chapter 1
# A New Science

It is difficult to say when exactly the computer was born. Some say a prototype was already there in the mechanism of Antikythera, realized in the third century BC, perhaps by Archimedes in Siracusa. Others state that it was created almost two thousand years later, in the seventeenth century, thanks to Wilhelm Schickard, or Blaise Pascal, or Gottfried Leibniz, who invented adding machines. Finally, some credit the realization of the first computer to Charles Babbage, who, at the beginning of the nineteenth century, designed a valuable analytical engine, although it was never actually produced. On the other hand, a good number of historians would say that the computer, the way we know it today, was an invention of the twentieth century and that, during its short life, it has gone through various stages of development: from the mechanical analog computer developed by Vannevar Bush in the USA in 1927, which could even solve a few differential equations, to the virtual computer proposed by Alan Turing in England in 1936, in order to solve the so-called "decision problem" of David Hilbert, down to the first electronic computer—ABC—developed by John Vincent Atanasoff and Clifford Berry in 1939 in the Physics Department of Iowa State University.

It goes without saying that the history of the developments that led to the birth of the modern computer has been far richer and more complex than the brief summary given above.[1] But it is enough for our present purposes. Electronics is the technology that makes the difference, since it allows unprecedented speed and processing power. Talking about machines allowing a qualitative change in the way science is produced—an epistemological leap—thanks to their incredible processing power, we can say that this is what happens with the modern computer launched between 1936 and 1939, with the virtual "Turing machine", and the prototype electronic machine of Atanasoff and Berry.

---

[1] For a comprehensive history of the development of the computer, see H.H. Goldstine, *The Computer from Pascal to von Neumann*, Princeton University Press 1972. Limited to the 20th century, but offering a picture not strictly confined to the United States, is the collection of essays edited by N. Metropolis, J. Howlett, and G.C. Rota, *A History of Computing in the Twentieth Century*, Academic Press 1980.

© Springer Nature Switzerland AG 2020
G. Battimelli et al., *Computer Meets Theoretical Physics*, The Frontiers Collection,
https://doi.org/10.1007/978-3-030-39399-1_1

Like all new inventions, the electronic computer did not immediately express its whole potential. After all, in Los Alamos, in the middle of the Second World War, with his adding machine, Enrico Fermi could compete in speed with John von Neumann and his computer. In short, from the 50's, electronic machines began to show a processing power incomparably greater than anything that could be achieved by man. Since then, computer processing power has increased by 18 orders of magnitude (i.e., a trillion times), thereby producing a profound transformation of scientific practice.

John von Neumann (1903–1957), Richard Feynman (1918–1988), Stan Ulam (1909–1984) at Los Alamos, circa 1945 (Los Alamos National Laboratory)

This is not the first instance of technology generating "new science". In fact, some philosophers say that technology and science stimulate each other's development. Indeed, science allows the development of new technologies. For instance, Einstein's general relativity provides the basis for the development of GPS technology, which enables satellites to move in space along certain trajectories, and allows our cell phones to indicate a precise route. But the reverse is also true: technology generates "new science". The eyepiece developed by Galileo Galilei in the summer of 1609 enabled him, not only to note that the Moon is the same kind of object as the Earth, but also that at least four moons orbit around Jupiter, and there are many more stars than those we can see with the naked eye. Moreover, Galileo made a huge epistemological leap and, in the words of Alexandre Koyré, passed «from the closed world to the infinite Universe».[2]

---

[2]A. Koyré, *From the Closed World to the Infinite Universe* (1957).

As we shall try to show in this book, the introduction and development of the electronic computer, with its unusual and ever-growing processing power, created the conditions for a scientific and epistemological change which can be compared, though on a smaller scale, to the one brought about by Galileo: namely, his "re-construction of Nature", which transformed him from a passive "legislator" into an active "constructor" of the world.

For the first time, thanks to the computer, it became possible to study the evolution in time of systems consisting of thousands—or even millions—of elements. This allowed scientists to simulate the behaviour of real macroscopic objects, and predict their properties. And so a new science was born, molecular simulation, of which molecular dynamics is the most accomplished achievement, whose protagonists try to leave behind, once again, a "closed world" and open the door to an "infinite Universe", where we have the potential to fully reconstruct the mechanisms explaining real macroscopic systems.[3] Indeed, with molecular dynamics one can "calculate theory", and hence simulate and predict the observable behaviour of real systems (including chemical and biological systems), by using as input only the laws of physics and its fundamental constants.

## 1.1 The Goal of Theoretical Physics

In principle, modern Science was born when humankind managed to shake off an idea of Aristotle, namely, the principle of impossibility, according to which man can only act "according to Nature". For example, Aristotle argued that automatic machines ("automatons") could not be built. For if such machines existed, we would no longer need slavery, and since slavery was "according to Nature", it could not be eliminated. As a consequence, automatons could not exist.

Aristotle's principle of impossibility came from a coherent, if restrictive, appli-cation of his principle of non-contradiction. This is an important principle. Indeed, the idea that a potential thing, contradicting something else that already has its own reality, cannot actually exist, is interesting and, in different contexts, it can even be useful. However, with regard to automatic machines, and to Nature in general, Aris-totle's idea defines false theoretical impossibilities, which may even be obnoxious for human development.

The English philosopher Francis Bacon, who lived between the end of the six-teenth and the beginning of the seventeenth century, was the first, clear-headed prophet of modern science. Bacon did not actually contribute to the development of modern science, but guessed with astonishing insight what research based on the

---

[3]The "new science" we mention should actually be defined as molecular simulation (MS), articulated into Metropolis Monte Carlo (MC) and molecular dynamics per se (MD). For the sake of simplicity, and in order to underline the fundamental scientific nature of MS, we shall sometimes use the particular for the general, by referring to MD instead of MS, when the topic of our discussion is clear from the context.

scientific method could do for humankind: in a word, it could master Nature, but under the strict condition of always submitting to its "laws".

Obviously, this is easier said than done. The question immediately becomes more involved when we try to define Nature's "laws". For how can we discover and/or invent them? What are their limits of validity? Here, Bacon's thinking becomes quite useless. However, we can learn a few good ideas from Galileo and, in general, from the concept of formal, or perhaps better, mathematical science of Nature.

In contrast to widespread opinion, Galileo's originality does not lie solely in his experimental method, in his insistence upon "sense experiences" – that is, upon empirical experiences, which have been realized by following a theoretical plan, "removing obstacles". Going beyond that, he had guessed the key role of "certain demonstrations": you must formulate all the phenomena studied with experiments in mathematical terms, so as to derive, through calculations, as many different kinds of behaviour as possible of the system under examination, from the minimum number of principles.

Galileo taught us to trust mathematics.

This is not due to the fact that mathematical formulae lead us to the truth, but rather to another fact, i.e., our trust in mathematics guarantees that, if a consequence is not correctly foreseen by the model we use to represent reality, we don't have to look around for external mistakes: we know that it is the model itself that is unsuitable. If the model does not correctly explain the experimental data, and we are using the mathematical procedure correctly, this means that we should change something in the model, starting from the more specific hypotheses and working down as necessary to the general principles from which the model is derived, with a rather linear and controlled procedure. And hence, mathematics plays a key role in the physical representation of the world.

It is then worthwhile to try and understand why this works. In particular, we should try to understand why mathematics is so effective in the natural sciences, and why it makes it easier to identify mistakes. Why is it so trustworthy, whenever it is applicable (because it is not always as easy as it is in physics, as we shall see later on). Once again, Galileo helps us with an intelligent and amusing explanation:

«Extensive, namely including the multitude of infinite intelligible objects, human understanding is practically null, even though it grasps a thousand propositions, since a thousand is zero in comparison with infinity. On the other hand, if we consider intensive – namely, perfect - understanding, I say that the human mind understands some of them so well, and is so absolutely certain of it, as if it had the same nature; such are pure mathematical sciences, namely Geometry and Arithmetics, of which the divine intellect understands infinitely more propositions, because it knows them. However, I think that the human knowledge of those few propositions is equal to divine knowledge because of their objective certainty, because it understands their necessity, about which there is an absolute confidence».[4]

---

[4]G. Galilei, *Dialogo sopra i due massimi sistemi del mondo Tolemaico e Copernicano* (Florence, Landini, 1632) II, p. 129. Translated by Stillman Drake, *Dialogue Concerning the Two World Systems - Ptolemaic and Copernican*, University of California Press 1967.

In other words, God—Galileo says—knows all possible and conceivable mathematics, therefore His *extensive* knowledge is infinitely greater than human knowledge, since He knows everything. On the other hand, we poor humans only know a few principles of mathematics and a few propositions about the world in the language of mathematics. However—Galileo adds—the little we do know, we know in an intensive way, just like God Himself, simply because mathematics is pure "truth". Divine truth. In modern, less theological terms, we can say that mathematics is a human product, rather than a natural product (external to man); therefore it has none of the margins of uncertainty normally associated with anything not made by man.

Summing up, the issues surrounding our knowledge of natural phenomena seem quite clear: if we manage to represent a wide range of phenomena accessible to experiment with a few equations, then those phenomena are wholly under our control, and we can consider ourselves their masters, according to Bacon's thinking. Indeed, we can explain well-known facts, foresee new ones, and use our mathematical knowledge to produce and exploit situations, different from those found in Nature, to our advantage.

Therefore, the problem of our scientific knowledge of the world becomes the following: when is it possible to sum up and represent the largest possible family of knowable phenomena—namely the Universe—with a few equations? For three hundred years, the exact sciences have tried to answer: always. In the human sciences, or in any case the farther away one moves from the exact sciences, the answer has been: never, or almost never, or, if we're optimistic, sometimes. The truth is less straightforward than that, and it is worthwhile to discuss this point at length.

Since the times of Isaac Newton, theoretical mechanics gained the respect of "practical" scientists, and this gave the impression that Bacon's idea was ready to hand. Thanks to reductionism (the atomic hypothesis and its by-products, whereby complex realities are but manifestations of atoms and their motions), and to the generalization of classical mechanics to quantum (and eventually relativistic) mechanics, even chemistry gradually took the same direction. However, neither biology, nor the social sciences, were able to follow this approach. Why? Out of necessity, perhaps, due basically to our failure to dominate large and complex problems? Or rather by accident, due to some historical constraint on our development?

If we look into matters in detail, the first hypothesis seems unreasonably pessimistic, since it is based upon a presumed human failure, whereas the second hypothesis seems much more convincing. Indeed, at least with regard to chemical technologies and biology, there seem to be clear signs of a deep transformation in these disciplines: both theoretical chemistry and theoretical biology are increasingly coming to resemble the best theoretical physics. We still find ourselves in an indefinite condition. No hypothesis can be fully demonstrated in the field of mathematized chemistry and, even less, biology. We are still waiting for convincing and, if possible, definitive proofs of all that. However, recent developments tend to reinforce the idea that the phenomena studied by both chemistry and biology will soon be under control, just like those studied by mechanics.

In other words, the dream of theoretical physics—to reconstruct the world—seems closer in contexts which were above suspicion until yesterday, such as the chemistry of materials and biology. Therefore, the best we can do is to accept the optimistic hypothesis and try to understand what this process has triggered, and what are the theoretical limitations that we are attempting to overcome.

The idea we shall try to demonstrate in this book is that the missing ingredient for the development of theoretical physics according to Bacon, in the last three centuries, has simply been calculating power. Nowadays, endowed with a certainly much refined Newtonian theory (although fundamentally proceeding along the same path), and with the increasing processing power made possible by a technological innovation—the computer—we may hope to realize Francis Bacon's dream: that is, get to know Nature so well, that we may think, without arrogance, that *Homo sapiens* will gain ascendancy over it.

It is easy to understand what we mean when we say that *Homo sapiens* may gain ascendancy over Nature. Indeed, since the seventeenth century, thanks to the research work leading to Newton's synthesis, the idea of science as "predictive knowledge and power over the world" has been virtually realized in classical mechanics, which has thus represented the fullest model of complete modern knowledge.

Mechanics has had significant results in predicting phenomena. For example, the discovery of Neptune in September 1846, made by Johann Gottfried Galle and one of his students, Heinrich Ludwig d'Arrest, at the Astronomical Observatory of Berlin, was a dramatic illustration of this. The two scientists pointed their telescope just where the Englishman John Couch Adams and the Frenchman Urbain Le Verrier had predicted, by strict application of the theory, that there would be an unknown planet. They had even guessed its mass, as confirmed by the scientists who found it. That was probably the moment in which the theory of mechanics gained its peak of popularity. In any case, whether it be ballistics, astronomy, or practical mechanics, the common feeling among those physicists was that they had got hold of a theoretical tool that would be able to solve all difficulties.

However, the ability to predict (relatively simple) phenomena has rarely been accompanied by any possibility of gaining ascendancy over Nature, as Francis Bacon had imagined. Knowledge was transformed into power over Nature only in peculiar contexts. The construction of instruments exploiting the laws of mechanics, or the application of astronomical knowledge in order to effectively navigate, are good examples. However, for a long time, scientists did not go much farther than that.

Things reached a different level as soon as scientists tried to go beyond the scope of purely mechanical phenomena. So much so that, at the beginning of the nineteenth century, the French sociologist and philosopher Auguste Comte, who is considered the founder of positivism, felt the need to classify science, not as a whole, but rather by defining each branch of science by its own principle and mode: thus thermology, optics, and, even more so, chemistry and the life sciences were introduced as separate from mechanics.

The eventual return of acoustic, electromagnetic, and optical phenomena within the domain of mechanics—whether in the discontinuous area (particles) or in the continuous area (classical field theory)—required more or less a century. In the meantime, just to get that far, the laws of mechanics had to be revised (relativity and, above all, quantum mechanics show this). However, this revision was not as deep as is often suggested, and did not so much concern the whole scientific explanation of the Universe, as the internal consistency of a few prerequisites.

More difficult was the recovery of the unity of mechanics and thermodynamics, achieved by the end of the nineteenth century with the help of new and conceptually surprising statistical methods which made it possible to understand macroscopic behaviour on the basis of purely mechanical systems. From this moment on, the unifying ambition of physics was no longer just a reasonable chimera that could only be demonstrated in restricted domains, but became an actual possibility, both in theory and application.

The really important novelty that came to complete the mechanistic ideal of both Galilei and Newton was the great achievement of Ludwig Boltzmann, who consistently used statistical methods in mechanics in order to obtain a single explanation of both macroscopic systems (among them, the Universe itself) and mechanical microscopic systems, i.e., "atoms". To support this aim, a new mathematical discipline was introduced and formalized: the theory of probability, whose axiomatic formulation was given by Andrej Nikolaevič Kolmogorov in 1933, just as the new quantum mechanics was being devised by Werner Heisenberg, Erwin Schrödinger, Wolfgang Pauli, and Paul Dirac, just to mention a few well-known names.

At this point, the explanatory and predictive properties of the physical theory were extended, and, in contrast to the general trend at the beginning of the nineteenth century, they involved not only all the so-called physical phenomena (i.e., mechanical, thermal, electromagnetic, etc.), but also the simpler phenomena of chemistry, and, at least in theory, even the basic phenomena of life.

## 1.2 The Development of Theoretical Physics and Computing

Although in technical terms this historical development was only completed at the beginning of the 30's, from a general point of view, the inception of this process had already been very clear to the greatest scientists working at the end of the nineteenth century. In this regard, the predictions made by the French mathematician Henri Poincaré on the future of physics, written at the dawn of the twentieth century, are incredibly far-sighted and accurate.[5]

However, there is an all important problem here. In order to understand its scale, we need to make the distinction between the virtual and actual predictive capacity of physical theory. The former is huge, but obviously cannot be satisfying in itself. Not everything that theory can predict in principle can actually be predicted in practice. The latter, namely actual predictive capacity, is minimal, all things considered. The difference between these two skills—which prevents virtual predictive capacity

---

[5]H. Poincaré, *La valeur de la science*, Flammarion, Paris 1890, especially Chaps. VII and VIII.

from becoming real—is just the lack of computing and processing power. If we had infinite computing power, the difference between the virtual and real possibilities for making predictions on the basis of the physical theory would become nothing (if we concede the completeness and universality of fundamental physical laws, in any event perfectible).

The limitations of computing and processing power are no small matter. Rather, they constitute a decisive question, which must be treated with the utmost respect. Indeed, if we understand this problem, we are led to an important epistemological change: scientific progress no longer consists, in this perspective, in adding new laws, but rather in deriving all possible consequences from the laws we already know. Only in the case of a defect—for example, if the results we obtain do not comply with empirical data—would we have to update well-known scientific laws. It follows that the effort (and the value) of research is no longer focused upon the creative activity of formulating laws, but rather upon the work of those who derive their consequences. In other words, it is focused upon the ability to rationally reconstruct a framework through an algorithmic derivation from simple general laws. In this activity, over and above counting upon remarkable computing power, the scientist must be creative, since he must find the algorithm through which the mathematical problem can be solved within a given computing power.

One well-known theoretical physicist, Paul Dirac, seems to have totally missed the point when he said that the task of theoretical physics consists in finding the laws, while "all the rest is chemistry".[6] However, in this way, Dirac practically destroyed Bacon's modern concept of science, namely, the capacity to derive all phenomena from the laws of physics, no matter how complex those phenomena may be. Chemistry represents the first and the clearest example of these phenomena.

Nowadays, we are in a new situation. Computing and processing power are significantly affecting the development of science, and the accompanying Baconian dream. Two extreme examples will allow us to clarify this point: the first is represented by the book *What is life?* which, with some arrogance, Erwin Schrödinger published in 1944, while the second comes from a remark about weather forecasting made by John von Neumann at the end of the 40's.

In *What is life?*, Erwin Schrödinger, quite rightly proud of the contribution of quantum mechanics to the development of our knowledge of the world, decided to discuss the importance of the new physics for understanding the phenomena of life. To some extent, this book is still extremely interesting, although we should say that by now it appears totally outdated, and in any case it does not represent a reference for biologists. Indeed, Schrödinger's problem did not consist in explaining in detail (or rather, calculating) the phenomena of life which he wanted to discuss, but rather in demonstrating that they were not incompatible with the established laws of physics. In fact, according to Schrödinger, physics can at best allow us to

---

[6]This claim may be apocryphal, i.e., Dirac may not actually have stated that, once the fundamental equations of quantum mechanics, which rule the behaviour of atomic systems, had been laid down, the main task was done, and the rest would be "chemistry" (a term which is used here to indicate an activity on a lower epistemological level). However, it has become part of the folklore of the physics community, and is therefore used here only to illustrate the point, with no claim to historical rigour.

guess certain biological regularities (or laws, according to Comte), without going any further! In other words, this is a step backwards in comparison with the Newtonian realization of the Baconian programme.

A completely different example is provided by John von Neumann, considered one of the founders of computational physics, and even of computing itself. This mathematician, originally from Hungary, had discovered and studied a few instabilities in the numerical solution of partial differential equations, and had then decided to apply his algorithms to the equations used in weather forecasting. His algorithms were at the time (and still are) "computer intensive", namely they require a huge computing power, and therefore the use of electronic devices. Von Neumann wanted to predict, on the basis of well-known physical laws, the evolution of the weather over a couple of hours. Once he had completed his check, the calculation was effectively launched on the computer. Unfortunately, it took the computer two days to "predict" the weather which should have occurred 46 h earlier. It was a prediction of the past rather than the future. Thus it was clear to von Neumann, as well as to many other scientists, that one should either discover more efficient algorithms or increase the computer's speed. Otherwise, the capacity of virtual prediction contained in physical laws would retain no practical interest, with all due respect to Dirac. We all know how the situation then evolved: algorithms were dramatically improved, and computers became much more powerful. Thus, nowadays, the situation has been reversed: in a couple of hours, we can produce a prediction valid for 48 h. And we keep improving.

The example of von Neumann, though apparently negative, helps to clarify the first important novelty introduced by computers. The first advantage of being able to calculate, i.e., in a *computer intensive* way, not manually, consists in the possibility of seeing the behaviour of the solutions of what are known as non-linear equations, whose qualitative behaviour we cannot really assess, since we have no analytic, i.e., exact, description of them. The first impact of the numerical capacity brought to theory by computers is simply this: we can study the behaviour of systems whose evolution obeys laws we can write down, but not solve. This is typically the case in physics, and a fortiori in mechanics. In this context, the speed and power of computers, thus the possibility of making massive calculations, has produced a great new divide.

We can further illustrate this divide, which we define as fundamental computer simulation, by comparing the way in which Cornelius Lanczos introduced mechanics in his book *The Variational Principles of Mechanics* (1949), a book that is still used and has valid contents, with the way it is introduced in any contemporary textbook. Lanczos presented in an elegant, intelligent, and profound manner the principles of classical mechanics, as they were developed over the course of several centuries. It is a useful and complete book, which tells us a lot about the different, but basically equivalent ways in which mechanical problems can be formulated, and finally solved. On the other hand, the goal of any modern textbook on mechanics consists in learning to "classify" solutions, i.e., it aims to provide an in-depth discussion of the possible behaviours of different systems, check their stability or instability, and identify criteria that will give clues about the behaviour of a given mechanical system. In this context, "numerics" is an essential component, and any ruse that can help to solve the problem is welcome.

Numerical computing is the branch of mathematical analysis concerned with the search for algorithms able to obtain numerical solutions to problems which cannot be solved analytically, that is, problems whose solutions cannot be obtained in an exact and definitive manner with paper and pencil. A numerical solution cannot be obtained in an absolutely correct manner in a finite time, and in general it cannot be obtained with paper and pencil alone. However, the speed and power of computers can be harnessed to find results as close as one wants to the exact solution. In a numerical calculation, therefore, we need to find a good idea, to develop a fast and effective algorithm.

Maniac, the electronic computer designed by John von Neumann and built by Nick Metropolis at Los Alamos in 1952 (Los Alamos National Laboratory)

Mind you, the "formulation" of the problem remains absolutely necessary. Indeed, there is no "solution" without a correct "formulation". But it is important to realize that the emphasis has shifted today. The main thing now is not finding a new formulation which would enable us to better understand a particular problem, but rather being able to face all challenges by brute force as they say (namely through massive calculations), in order to draw all possible consequences, and finally, predict the system's behaviour effectively, rather than virtually, under given conditions.

If the results of the simulation differ from empirical data, then we clearly need a new formulation of the problem, as well as a new theory. This is yet another benefit of using simulations, of great epistemological importance.

We have reached the end of our general considerations on the current situation: the laws of physics, unless proved inadequate, are universally true. They are applied, and always valid, effectively rather than virtually. The world is made of space, time (that is, motion), and matter (that is, inertia as quantified by mass, and interactions). We do know the laws of evolution of the Universe. We now have to solve the equations they imply by finding the proper algorithms and calculating their consequences.

Let us mention Francis Bacon once again: we can only command Nature if we humbly accept to submit to its laws. While in the past it was fundamental to find the laws of Nature, nowadays there is a different problem. Indeed, we do not have to find new laws of Nature. Rather, thanks to the processing power offered by computers, we can concentrate on calculating the consequences of those laws. In other words, we can demonstrate the capacity of those laws to solve problems. Therefore, the current challenge is to invent new, more powerful processors, new algorithms, and new approaches. This is the most interesting part of the procedure we are trying to describe. We shall indeed focus upon this part in the last section of our discussion.

The success of the programme we have been trying to illustrate so far—that is, the success of computer simulation—is based upon a feature of physics which greatly simplifies the solution of the reductionist issue we have raised, i.e., the mathematical reconstruction of the Universe starting from the ultimate and simplest elements. Here, we could use the word "atoms" in its philosophical, rather than physical meaning, because today's atoms are by no means the ultimate elements. Indeed, the "ultimate elements" are now the elementary particles. The point is that, as the object we want to explain becomes more complex, its "atoms" grow into larger structures, while maintaining the simplicity which distinguishes atoms.

Let us try to explain this idea better. Chemistry is built up from atoms, first forming molecules, then molecular clusters and aggregates. The discovery that an atom is not a simple body, but made up of electrons and a nucleus, and that the latter itself is not elementary, but can be "broken" at leisure into fragments called protons and neutrons, and that these in turn are made up of quarks, has not changed the situation of chemistry. In fact, nuclear physics more or less left things as they were for chemistry, since the main thing for the chemist is the electromagnetic interaction between nuclei and electrons, considered as stable bodies. The complexity of the nucleus, though real, is not called into question, unless one wishes to take it into consideration. In the latter case, there are other fundamental interactions (strong and weak) to be considered, and the whole specific framework of particle physics. And incidentally, the specific framework is the same, apart from possible simplifying approximations which can be used at low energy, when the nucleus is stable. That is why we can solve one problem at a time, and do not have to take into account all levels of complexity at the same time. For example, in order to solve the problem of gauging the stability and properties of atoms, we must consider at least their nuclei and electrons, but we may simply consider them as material points endowed with mass, charge, and magnetic moment, and otherwise suppose they have no inner structure, even though we know that this is not true under any conditions.

In the same way, if we want to build molecules, we do not need to consider together all nuclei and all electrons. Indeed, we would simply consider heavy material points, constituted by nuclei and electrons linked to them, and other light material points, constituted by the remaining electrons. The two groups of objects will thus emerge as "elementary" within the molecular problem we wish to solve.

In short, we can say that, in order to solve problems, the chemist does not need nuclear physics, even though the nucleus of any atom is a real body. On the other hand, nuclear physics plays a major role when we are concerned with the subatomic world. Furthermore, if we wish to put together several molecules and study the behaviour of a molecular aggregate (i.e., ordinary matter), we should simply consider the molecules in play as material points, or clusters of points interacting among themselves according to given laws.

As the reader may have guessed, the game can go on forever. All that matters is knowing that the laws ruling these "elementary" objects (considered elementary in that part of the world we choose to observe) are always known: they are the laws of mechanics, transposed from one space-time scale to another.

This situation was already well-known in the 30's, before electronic computers came on the scene, with their speed and huge processing power. But we were only able to exploit the full potential of this situation on a large scale after the Second World War, when computers became available. Before that, the best one could do was to invent very simple mathematical models, containing the essential information, about a few levels of complexity of matter. But if one is interested in ordinary matter, one should not only solve more realistic models. One should, above all, systematically calculate the essential information along a progressive path: from the level of atoms up to molecules (studying them as interacting material points), and to more and more complex clusters of molecules (hard and soft matter, etc.). In this way, one is no longer engaged in what physicists refer to with some arrogance as phenomenology, but rather predictive theory, ennobled by the Latin phrase ab initio.

A cultural degeneration of this situation, on which we shall not dwell, is the establishment of unrestrained mathematical modelling for any kind of phenomena, even in contexts which cannot really be adapted to mathematics, such as the social sciences, or the explanation of history (which someone, with a suitable metaphor, referred to as "doctrines informes").[7] As a consequence of this trend, one often sees defined as scientific what really are worthless mathematical games.

Let us get back to our main topic. In the last sixty years, the success of this programme has been explosive, and has affected not only the *ab initio* prediction of well-known phenomena, but also the discovery of new phenomena through the ever more skilful management of high-performance experimental equipment: suffice it to say that experimentally we can now "see" space structures resolved down to atomic dimensions. As a result, we have learnt to predict the structures of molecules that include not just a few atoms, but rather thousands of them, and the physical properties of natural or synthetic materials, with the molecular complexity of polymers. Slowly but surely, we are realizing that the biological is no more complex than that, so we may hope that, in our generation, we shall be able to show in a conclusive rather

---

[7]G. Canguilhem (ed.), *La Mathématisation des doctrines informes*, Hermann 1972.

than merely virtual way that even life can be built up (not simply understood!) from matter and motion, that is, using the usual instruments of physics.

However, this is not Comte's physics, but a discipline which is no longer solely concerned with physical phenomena (as is often stated to define physics, in a rather circular manner, in school textbooks), and indeed a science that has permeated both chemistry and biology. On the other hand, it has become more and more difficult to define a physicist, inasmuch as physics, being everywhere, will no longer be characterized as a discipline alongside others, but as the science which can build the whole world, not just understand it. We can describe this process as the numerical resolution of complexity, starting from fundamental laws. In short, fundamental computer simulation.

The pathway to simulation has not been an easy one. On the contrary, in the beginning, the discipline had a hard time establishing itself as a new way of making physical theory. Nowadays, that is no longer the case. Computer simulation is a daily part of the work of many theoretical physicists, maybe most of them. On the other hand, as Max Planck used to say, in physics a novelty is only accepted when the generation used to old mental habits disappears and is replaced by a new one, grown up with new mental habits. Nowadays, the new generations of physicists have realized the great opportunities offered by the processing power of contemporary computers.

# Chapter 2
# The Origins of Simulation

*Phase Transition for a Hard Sphere System.*[1] This title is likely to be rather mean-ingless to laypersons. It certainly does not seem to announce a revolution in physics. Even the first line of the short paper—no more than two columns long—which was published in *The Journal of Chemical Physics*, and was written with the aseptic style so typical of scientific reports, is not particularly significant for those who know very little about thermodynamics: «A calculation of molecular dynamic motion has been designed principally to study the relaxations accompanying various non-equilibrium phenomena».

But going to the second sentence, one can sense that the authors really do have new and ambitious goals: «The method consists of solving exactly the simultane-ous classical equations of motion of several hundred particles by means of fast electronic computers». With this paper and these few words, Berni Alder and Tom Wainwright—two young physicists working at the University of California Radiation Laboratory in Livermore, California—started a revolution in physics, made possible by the advent of a new technological tool: the electronic computer.

Indeed, with this note signed by Berni Alder and his collaborator Tom Wain-wright in the month of August 1957, molecular dynamics was born. It presented itself as a completely new approach which, thanks to the speed and power of a new technological tool—the electronic computer—offered the prospect of realizing one of physicists' longest standing dreams, namely «solving exactly the simultaneous classical equations of motion of several hundred particles». This meant simulating the dynamics of complex systems, composed of a large number of highly interacting components.

Alder and Wainwright were grappling with an issue which had been on the table for a few years: whether one may ascertain and correctly describe a phase transition, from the fluid to the solid state, within a system whose components behave like hard spheres. We shall soon go into this in some detail. But for the time being, let us simply say that the hard-sphere model is ideal, because the small balls studied by Alder and

---

[1]B.J. Alder, T.E. Wainwright, *Phase Transition for a Hard Sphere System*, Journal of Chemical Physics **27**, 1208, 1957.

© Springer Nature Switzerland AG 2020
G. Battimelli et al., *Computer Meets Theoretical Physics*, The Frontiers Collection,
https://doi.org/10.1007/978-3-030-39399-1_2

Wainwright can only collide and bounce back: they have no other interaction, unlike the atoms and molecules which compose real fluids and solids. Let us also add that, until the first half of the 50's, anybody who wanted to face the problem of phase transitions within a system composed of ideal spheres or real molecules, would face a barrier of impossibility. Using classical analytical methods, there is no way to solve the problem of finding a correct solution to the classical equations of simultaneous motion for several hundred particles, even in the case of ideal hard spheres.

Now, here is what young Alder and Wainwright stated, in their paper published in the *Journal of Chemical Physics* in the month of August 1957: we have overcome the barrier of impossibility, and have shown that it is finally possible to get to grips with the problem, because we have the unprecedented opportunity to simulate molecular dynamics. In other words, we can solve the equations of motion within a multi-body system with an exact numerical method, thanks to an algorithm which we have developed, and to the computing power of the new electronic computers, which can calculate much more quickly than any man.

That's how it works in science. A really innovative technology suddenly makes possible something which has so far remained out of reach. Let us mention for instance Galileo's eyepiece, which finally made it possible "to see things which had never been seen before", and inaugurate modern astronomy. The electronic computer is worthy of Galileo's eyepiece, since it finally enables scientists to realize Gottfried Leibniz's impossible dream: to have at one's disposal a machine which frees man from the hard work of calculation.

The German philosopher and mathematician had already imagined and almost triggered this new era, as in 1673 he had shown, both at the Académie des Sciences in Paris and at the Royal Society of London, a device known today as the "Leibniz wheel", capable of making linear operations (addition and subtraction), as well as multiplications and divisions. Being a philosopher, Leibniz had guessed the novelty of his original device: «Also the astronomers surely will not have to continue to exercise the patience which is required for computation. It is this that deters them from computing or correcting tables, from the construction of Ephemerides, from working on hypotheses, and from discussions of observations with each other. For it is unworthy of excellent men to lose hours like slaves in the labor of calculation which could safely be relegated to anyone else if machines were used».[2]

Perhaps Leibniz went too far. His "wheel" did not exempt either astronomers or other scientists from wasting hours like slaves on computing tasks. Nowadays, three hundred years on, the ambition of the great mathematician and philosopher is finally being realized. The electronic computer can free astronomers and many other scientists from the need to exercise patience in calculation. Thus, excellent men no longer need to waste hours like slaves on computing tasks. They have time to think and create.

---

[2]Quoted in H. Goldstine, *The Computer. From Pascal to von Neumann*, Princeton UP, Princeton, New Jersey, USA, 1993, p. 8.

In fact, the electronic computer goes far beyond Leibniz' dream. Not only does it free man from the hard work of computing, but it also flings the doors wide open to a new scientific method, namely simulation, thanks to a processing speed several million times higher than human speed.

This is what Alder and Wainwright did in the mid-50's. They thought and created. They also used the mechanical slave to propose a correct solution to their problem. At first sight, this may seem a marginal problem, only interesting to the initiated. Indeed, the system studied by these two researchers at the Livermore Laboratory was still ideal, and rather simple: it only involved 32, or at most 108 hard spheres. However, in just one hour, with their new computer, an IBM-704, they reproduced 7,000 virtual collisions in a 32 hard-sphere system and 2000 virtual collisions in a 108 hard-sphere system. In this way, they started a brand new and quite general method for the practice of physics: computer simulation.

Strictly speaking, we should say that molecular dynamics originated earlier on, with Alder's presentation at a conference in Brussels in 1956. This happened in August, during a large conference on the statistical mechanics of irreversible processes. The conference had been organized by Ilya Prigogine, a physical chemist of Russian origin, who taught at the Free University of Brussels, where he had been carrying out intensive and original research on processes which can take place only in one direction (for example, spontaneous heat transfer from a hot to a cold body) and not vice versa, something physicists and chemists refer to as "irreversible processes".

Prigogine was a rising star. Indeed, in 1959, he became director of the International Solvay Institute for Physics and Chemistry, in Brussels, and in 1977 he was awarded the Nobel Prize for his studies of systems far from thermodynamic equilibrium. And back in the month of August 1956, he was already a well-known scientist, capable of attracting to Brussels the world's best specialists on irreversible physical and chemical processes.

More than fifty scientists brought their contributions to the August meeting. Taken together, they covered a wide range of open questions regarding the statistical mechanics of irreversible processes. Not all contributions were so striking. However, one of them was remarkable first of all for its title: *Molecular Dynamics by Electronic Computers.* It was proposed by our two young American scientists from the Livermore Laboratories, Berni Alder and Tom Wainwright, and it demonstrated that one can obtain a new, advanced knowledge of the evolution in time of a multi-body system with the help of electronic computers.[3] In the details of their paper, Alder and Wainwright described the dynamics of that peculiar ideal system, composed of about 100 hard spheres which we mentioned above. In any case, it was clear to all present that a new perspective would be opened up to all those scientists who sought to reconstruct the dynamics of real, complex systems, as soon as fast and powerful electronic computers became available.

---

[3]B.J. Alder, T.E. Wainwright, *Molecular Dynamics by Electronic Computers*, in I. Prigogine (ed.), *Proceedings of the International Symposium on Transport Processes in Statistical Mechanics* (*Brussels, August 27–31, 1956*), Interscience Pub., London 1958, pp. 97–131.

Alder and Wainwright seem to be opening the floodgates, not only to a new field of study, but also to a brand new way of carrying out research in physics. In fact, we may say that, even before their paper was published in *The Journal of Chemical Physics*, molecular dynamics originated in Brussels in the month of August 1956.

## 2.1   Berni Alder in Pasadena

So who are these two young physicists announcing a revolutionary step in science? Berni Alder was not yet 32 years old, but he had already had a busy life behind him, quite apart from his studies. He was a Swiss citizen, even though he was born in Duisburg, Germany, on September 9, 1925. His father was a chemist, his mother a housewife. They were a Jewish family. Berni had two brothers, one of whom was his twin. His father, who obtained a doctorate in chemistry in Munich, could not embark on an academic career because of the heavy antisemitic climate in Germany at the time, and in Munich in particular. Unable to fulfil his lifelong dream and become a professor, Dr. Alder settled down and worked for a long time as a manager in a US firm, Aluminum of America. In any case, Berni's family were part of the upper middle-class scene in Duisburg.

However, when Adolf Hitler rose to power in Berlin, in the first few months of 1933, the Alder family, just like all the Jews in Germany, started to seriously fear for their lives. History would soon show that this fear was far from groundless. In any case, Aluminum of America went out of business, and doctor Alder had to find a new job. Thus he decided to move with his family to a safer country, Switzerland, where he chose to live in Zurich.

Small, neutral Switzerland was a relatively safe country. It had been free from attacks for years. However, it was still far too close to Germany, which, even in the recent past, in the course of the First World War, had invaded two other small neutral states with no scruples whatsoever, viz., Belgium and Luxembourg. And by this time, German militarism had grown even stronger and more aggressive than at the time of Kaiser Wilhelm II.

The spectre of war was haunting Europe. Moreover the Alders, just like many others on the Old Continent, were witnessing with dismay and growing concern the persecution of Jews carried out by Hitler in Germany, and, starting from 1938, even in Austria, which had been annexed without firing a shot. Rather, the Austrian people had applauded. The situation became even worse by the fall of 1938, when Mussolini's Italy approved the notorious racial laws. Switzerland began to look more and more like a small democratic *enclave*, surrounded by well-armed fascist regimes which were persecuting the Jewish people. Perhaps not even Switzerland was really a safe place any longer.

Therefore, when Germany invaded Poland on September 1, 1939, thereby initiating the Second World War, the Alders wondered whether it would be better for them to move on once again and settle overseas. The problem became more urgent still when the Wehrmacht occupied neutral Belgium and Luxembourg with impunity, and entered Paris on June 1, 1940. France finally capitulated on June 25.

Now Switzerland was completely surrounded by fascist regimes which flaunted and put into practice their antisemitism. The small Swiss Republic appeared fragile, in the eye of this storm. Indeed, Hitler's troops were amassed on the border with Switzerland. An invasion was feared. The family had to hurry up. Berni did not fully realize the situation. He was still a young and carefree child. His father, on the other hand, was all too aware of the imminent danger. In the name of the whole family, he thus requested a visa for the USA and obtained it two years later. In 1941 the Alders had to cross the whole of France, now occupied by the Nazis, inside a sealed train. They reached Spain and finally Portugal, where they embarked on a ship that sailed for the USA from Lisbon harbour.

It was not an easy journey. Indeed, the ship was blocked by German U-boats in the middle of the Ocean. Hitler's sailors came aboard in order to check all the passengers, one by one. Then they let them go: after all, they were not (yet) at war with the United States of America, nor with Portugal. The ship could resume its journey.

The Alders felt they were safe when they disembarked in New York, in the month of April 1941 and saw the Statue of Liberty. But even so, they did not settle on the East Coast. Rather, they decided to reach Berkeley, on the opposite coast, since their uncle was waiting for them over there.

Thus Berni Alder, at 16 years of age, arrived in California. At the time, he could not speak English. Within three months, he had already learnt the language, and the situation at school was better than he could have imagined. After all, he had received a good education in the Swiss secondary school he had attended, so he found it easy to settle down in the new US high school.

Before long, the United States suffered the attack of Pearl Harbour (on December 7, 1941) and, in its turn, entered the war. For the time being, Berni Alder was not really worried about this. Rather, he decided to follow in his father's footsteps, and began to study chemistry at the University of California in Berkeley. In fact, this choice was not really free, but more or less imposed by his father. His older brother, Davis, followed his own inclination and entered the faculty of mathematics. He would eventually become a pure mathematician. On the other hand, the twins followed their father's will. Actually, Berni's twin brother became a dentist in the end. However, they would all work there, around the bay of San Francisco, or not too far away.

Berni Alder around the mid-fifties

Let us return to Berni. As a young student of chemistry, he showed great promise. He got on very well with his studies, and received regular praise from his teachers, apart from the fact that he tended to break test tubes in the lab.[4] Perhaps, this evidence of poor manual skills indicated his lack of interest for organic chemistry, and even for all applied sciences. Berni was by now certain that he would not follow in his father's footsteps, since he realized that he was at heart a theorist.

Meanwhile, the war would soon claim its due. Berni was 18, and must therefore join the army. In fact, since he was still a Swiss citizen, he could choose to go back to his own country. On the other hand, if he wanted to stay in the USA, he had to take up the new citizenship and contribute to the defence of his adopted homeland. Berni's choice was to become an American. How could it have been otherwise?

But was it really more useful for the USA to send him to the front than keep him in a lab in its bid to wage war against Japan, Germany, and Italy? Two of his chemistry teachers, Wendell Latimer and Kenneth Pitzer, did not think so. They thus tried to get him into a special lab at Berkeley, managed by Glenn Seaborg, which was part of the secret Manhattan project, set up to design the most powerful weapon of mass destruction ever achieved, the atomic bomb.

But their attempts failed. Berni was signed up by the US Navy. Although still a Swiss citizen, he had professed his intention to become an American citizen. Thus he trained for a year in Chicago, Treasure Island, and Monterey, on the use of a

_____

[4]Mac Kernan et al., *Interview with Berni Alder*, SIMU Newsletter no. 4, 2002, https://doi.org/10.13140/2.1.2562.7843, https://www.researchgate.net/publication/267979976_ SIMU_Challenges_in_Molecular_Simulations_Bridging_the_Length_and_Timescales_gap_ Volume_4.

technology which would contribute more than anything else to the Allies' victory: radar. He was then sent to a communication station in the Philippine jungle, near Manila. In this mosquito-infested area, Berni made his small contribution to the creation of the most powerful radar station in the Philippines, where there were still many Japanese soldiers. This radar station would prove useful throughout the Pacific campaign, and would make a noteworthy contribution to the final Allied victory.

At the end of the War, Berni Alder was discharged. This meant that he could go back to Berkeley, where he completed his studies, and in 1946, he obtained the title of Bachelor of Science in chemistry. Within just one year, in 1947, Berni Alder also obtained a Master's degree in chemical engineering. However, he had not changed his mind: this was not the subject matter he was really interested in, since it was applied science. Alder was attracted more than ever by theoretical problems. He felt that Berkeley was not the right place to pursue those interests.

He talked about this with Joel Henry Hildebrand, a highly competent chemical physicist, who was an expert in fluids and was teaching at Berkeley: «I really liked Hildebrand», Alder said years later. «He helped me in my career and, of course, he got me interested in liquids. He made me think about the problems of liquids, but he really didn't help me theoretically, as he was really into thermodynamics ... You see, Berkeley had a very anti-theory atmosphere, so I couldn't stay on there, and had to go».

Indeed, Berni Alder left Berkeley, with his new interest for liquids. In fact, he did not have to go far. There is a distance of only 375 miles between Berkeley and Pasadena, home of the California Institute of Technology. Here there was a completely different atmosphere for the theoretical chemistry specialists. Among the staff of Caltech in Pasadena, we may mention Richard Tolman, a mathematician, physicist, and chemist, considered an authority on statistical mechanics. Not only that. During the Second World War, Tolman had been the scientific advisor of General Leslie Groves, director of the Manhattan Project, which had succeeded in building the atomic bomb. Soon after the War, Tolman was appointed advisor to Bernard Baruch, the politician who created "the Baruch plan", aimed at preventing the proliferation of nuclear weapons. Tolman was therefore a protagonist in this unprecedented common ground that brought together science, the military, and politics in the USA.

However, Berni Alder did not realize all this, all the more so because he had hardly arrived in Pasadena when Richard Tolman died, on September 5, 1948. Nevertheless, two of Tolman's heirs were working at Caltech. One of them was Linus Pauling, an accomplished scholar who had been a student of Tolman. Pauling was the world's major expert on quantum chemistry, and was beginning to turn his attention to biological molecules, such as haemoglobin. In 1954, he won the Nobel Prize in Chemistry, both for his research on the nature of chemical bonds, and for his studies on the structure of complex substances. In 1962, he obtained the Nobel Peace Prize. Pauling, just like Tolman, was quite at ease in the new post-academic science that was quickly taking hold in the USA and in the rest of the world. Moreover, Alder was impressed by another feature he noticed in Pauling: he was an excellent example of what would be called an intuitive scientist.

Berni Alder decided he would like to work with Pauling. Unfortunately, Pauling did not accept PhD students, although it looked as though he might make an exception with Berni. Indeed, he talked about this with another remarkable colleague at the faculty, John Gamble Kirkwood, and they decided that Berni Alder should work on his PhD thesis with Kirkwood.

Kirkwood was another highly competent chemical physicist who had just arrived in Pasadena. He had came from Cornell University, and was renowned for his extreme rigour, as well as for his notable contributions to the development of the electrophoresis of proteins, and above all for his elegant mathematical approach to the study of the electric properties of both gases and liquids.[5]

Kirkwood, who was an expert on statistical mechanics, was convinced that only a mathematical calculation which took into account molecular interactions might provide a way to really understand the properties of a liquid. However, he knew that this endeavour was analytically impossible. That is why, between the 30's and 40's, Kirkwood had suggested circumventing the obstacle by proceeding through approximations. These, he said, could only be intuitions. The first and best known of them was referred to as the "Kirkwood superposition approximation", used to calculate the radial distribution function of a realistic fluid model directly from the fundamental principles of physics, that is, in this specific case, of statistical mechanics. In other words, the idea was to solve the fundamental equations of a fluid system for the probability distribution functions of its molecules.

Thus, in Pasadena, Berni Alder started studying for his PhD under the guide of the "rigorous" Kirkwood, instead of the "intuitive" Pauling. While working on the problems posed by Kirkwood, the young student came across a computer. It was not a case of love at first sight. Alder was not particularly enamoured of the work Kirkwood suggested, and he did not really like this tool—the computer—with which he was supposed to carry it out. However, his initial antipathy soon turned into love.

## 2.2   Meet the Computer

Meanwhile, as we mentioned above, a brand-new machine had been created: the electronic computer. It was much faster than the electromechanical computer, which was slowed down by relays. Back in May 1943 the US army had entrusted John Mauchly and John Eckert of the University of Pennsylvania with the realization of ENIAC, an entirely electronic computer, on the basis of a report which the two scientists had submitted in the month of August 1942.

The project turned out well. Indeed, ENIAC became operational on February 15, 1946 at the shooting range of the Artillery Command of Aberdeen, Maryland. The research group, composed of at least thirty mathematicians and countless technicians, was directed by Eckert, Mauchly, and Captain Hermann H. Goldstine, who was the

---

[5]S.A. Rice, F.H. Stillinger, *John Gamble Kirkwood* (*1907–1959*). *A Biographical Memoir*, Biographical Memoirs, vol. 77, The National Academy Press, Washington, 1999, p. 6.

director of the army's centre for ballistic studies, and based at the University of Pennsylvania at Philadelphia. ENIAC's circuits were indeed electronic—there was no mechanical moving part in them.

ENIAC was an automatic computer. This meant that, once an external operator had specified a program and presented it to the computer through a series of punched cards, the machine carried it through on its own, without further human intervention. This first electronic computer was digital, and it processed information. John Tukey identified the basic information unit, namely the bit (binary digit), which described one of the two possible states: 1 or 0, on or off.

ENIAC was a real giant: it occupied a $9 \times 15$ square meter hall, weighed 30 tons, and employed 17,468 valves of 16 different types, which produced 100,000 impulses per second, along with 70,000 resistances and 10,000 condensers. When it was running, it absorbed 170 kW. And it cost a fortune. However, it constituted a major breakthrough: it could carry out 5,000 sums or 300 multiplications per second. This meant that it was more or less a thousand times faster than any other mechanical or electromechanical computer, such as the one available to Alder in Pasadena.

The reason why an electronic computer can be faster than an electromechanical one by several orders of magnitude is easy to explain. In the latter, a crucial speed element is the relay, where an electric impulse is transformed into a mechanic movement, which is completed within 1–10 ms. On the other hand, in an electronic computer, which only employs thermionic valves, the only moving elements are electrons, which travel almost at the speed of light. Therefore, calculations can in principle be made in almost negligible times.

With the electronic computer, Leibniz' dream—to free scientists from the hardships of calculation—had at last come true. This can be illustrated by the fact that it was used in 1948, not only for the ballistic calculations so dear to the army (it could do in 30 s what would take an electromechanical machine 20 h), but also in a scientific environment: research on cosmic rays, applications to nuclear energy (also for the construction of the H bomb), the planning of wind tunnels, and meteorological studies.

Meanwhile, in the history of computer development, we find John Janos von Neumann, one of the greatest mathematicians of the twentieth century. He was a pure mathematician, but did not despise the idea of applying mathematics to physical problems, such as fluid dynamics. In the midst of war, because of his skills and his genius, von Neumann was working on the Manhattan Project. He was engaged in a crucial aspect of this, namely the implosion that brings the fissile material into a nuclear chain reaction.

Having heard of the plan to build the first electronic computer, starting from August 1944, John von Neumann was engaged in designing a machine that could "overcome ENIAC". The product of this work is contained in a 101 page report, *First Draft of a Report on the EDVAC,* which the Hungarian scientist submitted to the members of Moore School on June 30, 1945, while, in Los Alamos, he was preparing the first test of the atomic bomb, which took place in Alamogordo on July 16.

Von Neumann's report is commonly considered to be the foundational text for the logic of modern computers. As such, in Goldstine's words, it is «the most important document ever written on computing and computers»[6]. Indeed, the Hungarian scientist suggested "the way" in which the architecture of the new electronic machines should be realized.

Summing up, in the second half of the 40's, there were both new generation computers, endowed with exceptional processing power, which would "free scientists from the hard work of calculation", and a new kind of logic which would allow everyone to use these machines. These features taken together opened the door to a new practice of science, simulation, which was not only new, but had never even been imagined before.

Let us therefore go back to Caltech in Pasadena, where, in 1948, Berni Alder started his PhD work. His dissertation consisted on the one hand in the development of Kirkwood's theory of conductivity, and on the other in the calculation of the radial distribution function of hard spheres, on the basis of an equation known as the Born-Yvon-Green-Kirkwood equation. In the second part of his dissertation, among other things, Alder had to check whether, according to the theory of liquids formulated by Kirkwood, a set of hard spheres would actually undergo a phase transition, passing from the fluid to the solid state. We are talking of a fluid state rather than a liquid state, because hard spheres can only collide and bounce off one another according to the laws of mechanics, and are not subjected to further interactions. In sum, they cannot form the peculiar state of matter which we call a liquid. Hard spheres can only form a fluid: that is, a sort of dense gas.

This was no minor study. Rather, it was a key component of Kirkwood's general strategy, which aimed to develop a theory of thermal conductivity. It was in fact a very complex theory, one that Alder would later judge rather sharply, saying that it was «totally wrong, painful and completely out of control». Kirkwood's improbable general theory was also shared by John Irving, a young PhD student from Princeton, on the East Coast, who had come to Pasadena to study statistical mechanics in depth. In particular, Irving was working on a paper with Kirkwood, on a theory of transport processes. Their goal was far-reaching: derive the equations of hydrodynamics from classical statistical mechanics.

Alder found it challenging to work with Irving, and remembers giving his own contribution to the work, which would be published in 1950, even though we can find no trace of Berni's collaboration in the published paper. Alder's help was not recognized. It does not really matter. The significance of this collaboration consists in the fact that Berni Alder developed the general idea of deriving the macroscopic properties of a fluid using a statistical approach, starting from the behaviour of the molecules composing the fluid itself. This idea came from the father of statistical mechanics, Ludwig Boltzmann, and would dominate all Alder's future research activity.

---

[6]H. Goldstine, *The computer. From Pascal to von Neumann*, Princeton UP, Princeton, New Jersey, USA, 1993, p. 191.

The study of a hard-sphere system constituted the second part of Alder's PhD dissertation. As already mentioned, he had to calculate the radial distribution function, that is, the way the density of matter varied around a certain point. There was already a series of mathematical tools for this calculation: the so-called integral-differential equations, which John Kirkwood had helped to formulate. Their analytic, i.e., exact solution was impossible, however. Therefore, they fell back on a kind of numerical integration, or approximate but accurate solutions, which involved a great deal of calculation.

At first, Alder started working with Eugene Maun, another of Kirkwood's young PhD students. However, the computing centre in Pasadena had no electronic computer available. They only had an IBM electromechanical machine, which was less powerful and slower than ENIAC or the computer that von Neumann was then building. On the other hand, even with this machine, Alder and Maun could confirm that the integral non-linear equation they were using to solve the problem of phase transitions in a hard-sphere system did not admit a unique solution. Indeed, the system soon faced the instabilities of the integral equations to be solved.

Alder and Maun used a Friedan Marchant electromechanical computer, and gathered the results in a paper which would be published later under the title *Radial distribution functions and the equation of state of a fluid composed of rigid spherical molecules*.[7] They could show that the distribution of hard spheres in the system was correlated with their density. Their solution was valid up to a crucial value, beyond which it lost its validity. This is what they wrote: «This value ... evidently represents the limit of stability of a fluid phase of rigid spheres. For greater densities, a crystalline phase is the stable phase. The transition between fluid and crystalline phases cannot be discussed quantitatively without an investigation of distribution functions in the crystalline phase itself. In the case of rigid spheres, it appears likely that the transition may be of second order rather than of first order, although at present this is no more than a surmise».

In fact, they were going too far. They had no chance of showing that the system would crystallize beyond the crucial value. The difference between the two phase transitions is that the first-order transition involves a discontinuity (this is the case for the transition from the solid to the liquid state), whereas the second-order transition takes place in a continuous manner (in fact, there is a discontinuity in the second derivative of the free energy). Let us not enter into the thermodynamic details of phase transitions. Let us simply say that Alder and Maun stated that the hard-sphere system would undergo a second-order phase transition. The problem was that they could not demonstrate this. However, the main thing for our story is not so much this first result, nor this sense of impossibility, as the fact that Alder had tested his first computational approach to the fundamental problems of statistical physics, and entered a field of study that would keep him busy for a long time, namely, phase transitions in hard-sphere systems.

---

[7]J.C. Kirkwood, E.K. Maun, B.J. Alder, *Radial Distribution Functions and the Equation of State of a Fluid Composed of Rigid Spherical Molecules*, Journal of Chemical Physics **18**, 1040, 1950.

Alder knew that, if he wanted to go on studying whether a system of hard spheres could show a transition from the fluid to the solid state, he had two options. The first had to be discarded, because Kirkwood's method would have to be applied, which would involve a huge quantity of calculations, and get only an approximate solution. The second option consisted in testing a new method. Unfortunately, this new direction of studies could not be continued with Maun, since Eugene fell ill and he had to find another partner. Alder went to talk to Stan Frankel, who had just arrived from Los Alamos to take charge of the computing centre recently set up at Caltech.

Stan Frankel (1919–1978) (Los Alamos National Laboratory)

Stanley Phillips—"Stan"—Frankel was around thirty years old at the time, born in Los Angeles in 1919. He had obtained a PhD in physics at the University of California in Berkeley. In 1942, he started working with Robert Oppenheimer at Berkeley. Together with Eldred Nelson, he also managed to develop good computational techniques to handle an ongoing problem, the diffusion of neutrons. Oppenheimer, an expert in the field of quantum mechanics, was soon appointed scientific director of the Manhattan project by General Leslie Groves, and moved to a secret location in the heart of New Mexico: Los Alamos. "The bomb" would be made down there. Oppen-heimer also invited Stan Frankel, co-opted into the theoretical division directed by Hans Bethe.

Frankel reached Los Alamos in 1943 and, together with Nelson, started working on an IBM electromechanical computer with punch cards. He was very good at this. By the end of the war, in the month of August 1945, he left Los Alamos to attend, together with Nicholas Metropolis, the Moore School of Engineering in Pennsylvania. The aim was to learn how to program the new, completely electronic, computers that were being produced there.

The main topic of research was the same from Los Alamos onwards: nuclear physics applied to the making of bombs. By the fall of 1945, Frankel and Metropolis had developed a very important algorithm for the production of atomic weapons. This was the same project on which Edward Teller, a physicist of Hungarian origin, had been working for some time.

However, Frankel did not stay in Pennsylvania for long. He decided to move to Caltech in Pasadena, and there he met Alder, who was still facing the challenge of hard-sphere physics. Berni asked him if they could work together on the problem, and Stan accepted, although this problem was very different from those he usually han-dled. Frankel was a computer whiz. At Los Alamos, Frankel had come into contact with the people who had devised the first Monte Carlo simulation techniques. Origi-nally created to speed up calculations of neutron properties in the context of nuclear weapon design, the method had been developed into a powerful tool to calculate physical quantities by taking averages over large sets of possible configurations of the system under study, obtained by a proper random sampling operation performed by the computer. Besides allowing Alder to make use of the more advanced IBM machines at Caltech, Frankel introduced him to the general idea of Monte Carlo, and together they developed their own approach, trying to apply the method to the hard-sphere problem.

## 2.3  Early Monte Carlo Simulations

There were already several simple realizations of this Monte Carlo method. We shall not tell the story of all these early realizations. The general idea had been established in Los Alamos just after the Second World War by Stan (Stanisław) Ulam, a mathematician of Polish origin. It was designed to calculate the diffusion of neutrons, a decisive problem for the realization of atomic weapons. Ulam had suggested using a statistical approach, which envisaged the creation of "random numbers" in order to obtain, to a certain approximation, the solution of problems that could not be solved by analytical methods. The approximate solution would get ever closer to the exact value as the statistical sample grew.

In fact, Enrico Fermi had invented a similar method in the 30's, when he was grappling with the problem of "slow neutron" scattering, these being the neutrons that were particularly efficient in provoking nuclear reactions. Fermi had published nothing about this. However, from the words of Emilio Segrè, one of the "young boys of via Panisperna" who would win the Nobel Prize for Physics later on, we know that, in order to realize his calculations on a statistical basis, Enrico Fermi had used a mechanical adding machine which he had designed on his own.[8]

The processing power of Fermi's mechanical computer was not exceptional. However, not even the electromechanical computers used in Los Alamos could fully exploit the possibilities of the statistical approach proposed by Ulam. As Nicholas Metropolis writes: «His remarkable mathematical background made Stan conscious of the fact that statistical sampling techniques could not really be used, because of the long and boring calculations»[9]. In short, by the mid-40's, even statistical methods were limited by the huge quantity of calculations they required.

The situation changed in the month of April 1946, when John von Neumann, Nicholas Metropolis, and Stan Frankel gave a lecture in Los Alamos to illustrate the results obtained in the study of nuclear fusion with ENIAC, the new electronic processor at the University of Pennsylvania. At this point, «Ulam realized that the speed of this new machine might enable the large-scale use of statistical sampling techniques in order to solve physical and mathematical problems».[10]

This was the breakthrough: electronic computers would finally allow scientists to use statistical methods. Indeed, with ENIAC, the method used by both Ulam and Fermi could be applied on a systematic basis. This use was so successful that, when the electronic computer was not working because it was moved from Philadelphia to Aberdeen in 1947, in order to continue his studies on neutron transport, Enrico Fermi was forced to design and build, together with Percy King, a small analogue computer, which they called FERMIAC.

In the same year, Stan Ulam once again suggested the use of statistical techniques for the study of neutron scattering, which had finally been made possible thanks to the processing power of the electronic computer. During a conversation with Nicholas Metropolis, these techniques were defined as the "Monte Carlo method", in honour of the well-known European casino, where statistical techniques were widely used by players in an intuitive rather than a rigorous manner. The name "Monte Carlo" officially appeared for the first time in the title of a paper published by Metropolis and Ulam in September 1949.[11]

---

[8]E. Segrè, *From X-ray to Quarks*, W.H. Freeman, San Francisco, 1980, p. 221.

[9]N. Metropolis, *The Beginning of the Monte Carlo Method*, Los Alamos Science, Special Issue, Los Alamos, 1987, pp. 125–130.

[10]W.W. Wood, *Early History of Computer Simulations in Statistical Mechanics*, in G. Ciccotti, W.G. Hoover (eds.), *Molecular-Dynamics Simulation of Statistical-Mechanical Systems*, Proceedings of the International School of Physics "Enrico Fermi", Course XCVII, North-Holland 1986, pp. 3–14.

[11]N. Metropolis, S. Ulam, *The Monte Carlo Method*, Journal of the American Statistical Association **44**, 335, 1949.

It should be noticed that both Ulam and Fermi before him applied their method to uniform samples, in which all states were equally probable. This is far from realistic if the idea is to apply the method to solve problems in statistical mechanics, where we have to deal with a (huge) set of states characterized by a highly irregular probability distribution. The refined version of the general Monte Carlo method that provided an effective way to address this problem was developed in the early 50's, and has since then been known universally as "Metropolis Monte Carlo", from the name of the first author of the paper in which it was presented, published in 1953.[12] This was the paper inaugurating simulation on a machine endowed with remarkable processing power, as well as a new technique for the solution of problems posed by statistical mechanics. Metropolis Monte Carlo was indeed the first effective method of computer simulation.

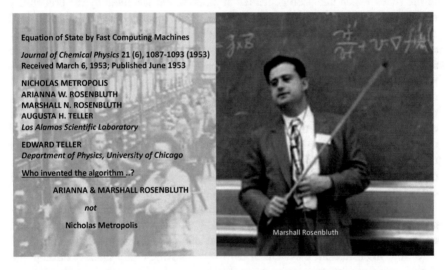

A slide taken from a recent presentation by Michael Klein, giving proper credit to the creators of the "Metropolis algorithm" (M. Klein)

In statistical mechanics, the thermodynamic properties of a group of particles, such as temperature or pressure, may in principle be calculated using suitable averaging operations on the possible microscopic states of the system (physicists refer to them as "points in phase space"). Let us take, for example, a system composed of 100 atoms inside a box. The phase space of this classical system is constituted of the position and velocity of each atom (which is treated as a material point for simplicity). We therefore have a wide range of possible physical states, since each atom may assume different positions and move with a different velocity. Not all states are equivalent, however. Indeed, each phase space point is associated with a "probability density", that is to say a mathematical function which allows us to calculate the likelihood that a given particle may be found in a given position and have a given velocity.

---

[12]N. Metropolis, A.W. Rosenbluth, M.N. Rosenbluth, A.H. Teller, E. Teller, *Equation of State Calculations by Fast Computing Machines*, Journal of Chemical Physics **21**, 1087, 1953.

Each general macroscopic property of our system can then be represented by a "function on the space of states". In our case, if we wish to determine the general property "temperature" of our system of 100 atoms, we must calculate the average kinetic energy of the system itself, "weighted" by the probability density of the assumed values. Statistics specialists call this "weighted average" an "expected value".

Normally, the microscopic property whose average value we wish to calculate (in our case, the average kinetic energy of the atoms), and hence the corresponding thermodynamic property (in our case, the temperature), is represented by a relatively regular function upon the phase space, whereas the probability is a highly irregular function.

The basic idea of the Monte Carlo method is to calculate the average on a sample of the states which has been generated at random, but weighted by the probability distribution of the *ensemble* being tested: the moves leading to a state characterized by a relatively low probability will be penalized in comparison to those leading to more probable situations. To put it simply, the method works like this: one starts from the first state, then proceeds by sampling new states, while following the instructions which, on a case-by-case basis, say whether to accept or reject the new state which is being proposed. If the procedure is defined correctly, and the iteration of the process is prolonged sufficiently, the states thus generated will asymptotically produce a sample of equilibrium states of the system, whence scientists may calculate the average on the *ensemble* as an arithmetic average over the sample, and thereby obtain its macroscopic thermodynamic properties. It can be demonstrated that the values thus obtained are very close to the real ones. More precisely, there are strict mathematical criteria which allow scientists to check the reliability of their results.

This in outline is what is known as the Metropolis Monte Carlo method[13]. Independently of the Los Alamos group, Alder and Frankel in Pasadena invented, between 1950 and 1951, essentially the same method and applied it to the particular model they were interested in: a system of hard spheres which only interact through the elastic collisions that take place when they impact. For the rest, the spheres move freely, without being affected by any other force. In this case, the difficulties of the Monte Carlo sampling technique drop markedly, because the rule of acceptance or exclusion for a new randomly generated configuration is simply to accept all situations, except those in which two spheres are superimposed. In this case, one goes back to square one, which is taken into account in the final count of states upon which the system's average properties are calculated.

---

[13] A detailed historical reconstruction of the process that led to the 1953 collective paper suggests that the main contribution to the work came from Marshall Rosenbluth, rather than from Nicolas Metropolis. However, the locution "Metropolis Monte Carlo" has by now been firmly established in the literature. On this point, see J.E. Gubernatis, *Marshall Rosenbluth and the Metropolis algorithm*, Physics of Plasmas **12**, 057303-1, 2005.

Daan Frenkel, Marshall Rosenbluth and David Ceperley at the conference celebrating the 50th anniversary of the Metropolis algorithm, held at Los Alamos in June, 2003 (D. Frenkel)

In spite of this early start, the results of the work done by the two scientists in Pasadena were published two years after the publication of the paper in which the Los Alamos group announced that they had developed the method which would eventually become known as Metropolis Monte Carlo.[14] Alder and Frankel published their paper, *Radial distribution function calculated by the Monte Carlo method for a hard sphere fluid*, together with Victor Lewinson, only in 1955.[15] The results were rather poor, and the publication arrived too late to be able to claim priority or, at least, concurrent discovery. Later on, in an interview granted to George Michael, Berni Alder claimed that he and Frankel had not used, but rather discovered, the new Monte Carlo method.[16] And indeed, in a footnote to their 1953 paper, the Los Alamos physicists give credit to Berni Alder and Stan Frankel for reaching the same results independently. At least the two young scientists received the honours of war.

---

[14]From now on, for the sake of simplicity and unless confusion can be generated, we will just use "Monte Carlo" for what should properly be referred to as "Metropolis Monte Carlo".

[15]B.J. Alder, S. Frankel, V. Lewinson, *Radial Distribution Function Calculated by the Monte Carlo Method for a Hard Sphere Fluid*, Journal of Chemical Physics **23**, 417, 1955.

[16]G. Michael, *An Interview with Bernie Alder. Stories of the Development of Large Scale Scientific Computing at Lawrence Livermore National Laboratory*, http://www.computer-history.info/Page1.dir/pages/Alder.html. It should be noted that the approach proposed by Alder and Frankel was correct, but only for hard spheres, as they didn't put forward any general criterion to treat acceptance. That would become essential when extending the method to models containing continuous potentials.

The reasons for this delay in publication were twofold: one was connected to the hardware (the still limited power of the machines then available, implying insufficient statistics and unreliable results), while the second had to do with Kirkwood's doubts about the theoretical foundations of the method.

Indeed, the statistics available to Berni and Stan was too limited. It was not sufficient to actually show anything, let alone that a hard-sphere system could undergo a phase transition. The problem was that at Caltech the chances of realizing this project were wholly virtual. The IBM electromechanical computer which Alder and Frankel were using did not have adequate processing power to fully exploit the new algorithm, not even for the relatively simple model of the hard-sphere system Alder was interested in. To solve that problem, and possibly convince Kirkwood of the validity of the algorithm, they would have needed a computing power that was not available in Pasadena. Although they immediately realized that «for the first time, it was finally possible to solve the many-body problem numerically», their results were clearly unreliable.

The situation changed in the summer of 1951, when Stan Frankel moved to Manchester, England, and had the opportunity to use the Ferranti MARK 1, the first electronic computer on the market, available at Manchester University. Frankel spent the whole summer trying to get the machine to solve the hard-sphere problem using the new Monte Carlo method, developed with Berni Alder. Frankel was glowing on his return to California. He explained to Berni that he had obtained good results with "their" algorithm. But the new "good results" were still not enough to convince Kirkwood of the validity of their method.

For quite some time, indeed, Kirkwood did not believe in their algorithm, convinced that it was ill-conceived. He had serious doubts about its validity on statistical grounds. Those doubts were dissipated later on, thanks mainly to the work of his student William Wayne (Bill) Wood, who proved the reliability of the probabilistic foundations of the method. But by the mid-50's Kirkwood was still sceptical. And, since he was the principal investigator of the project, he could not therefore endorse its results. Alder and Frankel could have challenged his authority, and publish their paper contrary to John Kirkwood's recommendations. However, they did not do this. «You cannot publish an idea your boss does not believe in», Alder stated. So the two young scientists missed the chance to launch, in their particular case, the Alder-Frankel Monte Carlo method, and were overtaken by Metropolis Monte Carlo.[17]

---

[17]There was, in fact, a remarkable difference: Alder and Frankel discovered the right method (for the very specific case of the hard-sphere model) without actually knowing why it worked, whereas the Los Alamos group fully realized this both from the physical and mathematical points of view. The fact is that the method became known only as the Metropolis Monte Carlo method.

Bill Wood (1924–2005) (Physics Today)

Kirkwood probably contacted the group in Los Alamos as well, asking for information. He knew they were working on a similar project, and that they were using a powerful electronic computer, the MANIAC. He knew that even there, in spite of the availability of the most powerful computer, people could not get convincing results on the matter of phase transitions. In 1954, the Rosenbluths published a paper that presented further interesting results but seemed to leave little hope that Monte Carlo would ever be useful for the specific problem of phase transitions: «Results obtained thus far lead us to feel strongly that the Monte Carlo method is a useful tool for solving statistical mechanical problems, although it does not appear to be feasible to obtain detailed results in transition regions».[18] This was not the sort of statement that would make Kirkwood more sympathetic to the method.

---

[18]M.N. Rosenbluth, A.W. Rosenbluth, *Further Results on Monte Carlo Equations of State*, Journal of Chemical Physics **22**, 881, 1954.

Meanwhile, already in 1951, just after discussing his dissertation, Berni Alder had left Pasadena to go back to Berkeley, where he got the chance to teach chemistry at the university for three years. Later on, between 1954 and 1955, he spent a whole year in Europe, including six months in Leiden, Netherlands, and six months in Cambridge, England, thanks to a Guggenheim fellowship.

The year spent in Europe was intensive, and further contributed to the education of the young Swiss scientist, who had since become an American citizen. Leiden was indeed a good centre for statistical mechanics, thanks to two physicists who had established and developed non-equilibrium thermodynamics: Peter Mazur and Sybren de Groot. Physicists skilled in quantum mechanics, such as George Eugene Uhlenbeck and Wolfgang Pauli, were also there. «Yes, I remember that it was a very active centre ... there was a good vibe», Alder said. At Cambridge University, Berni first worked with John Lennard-Jones, a theoretical physicist and chemist. He subsequently worked with the physicist Stephen Pople and the chemist Hugh Christopher Longuet-Higgins. «It was a very good time», Alder said of his semester in Cambridge.

## 2.4   The Birth of Molecular Dynamics

In the framework of the Manhattan Project in Los Alamos, Edward Teller led a group of theorists who studied nuclear fusion, rather than fission. This was still his main research goal: to build new, more powerful, and more effective nuclear weapons. The Cold War was in full swing. The United States thought that the new nuclear weapon would not only be useful, but also necessary. The Administration in Washington generously funded research on new weapons of mass destruction. Thus, in 1952, Teller established a large, fully equipped research centre at Livermore, in order to be able to continue the studies he had started at Los Alamos on nuclear weapons in general, and on fusion weapons in particular. So the Lawrence Livermore National Laboratory was established. Edward Teller was its associate director.

Teller immediately asked Alder to work with him and become an advisor of the new research group which was being formed in Livermore. This was an implicit acknowledgement of his worth. The distance between Berkeley and Livermore is barely 60 kilometres, so it was not impossible for Berni Alder to continue to teach chemistry at Berkeley and carry on his research at Livermore. He therefore joined Teller's team in this new research centre, run by the University of California, whose mission was not to replace, but rather to extend the Los Alamos National Laboratories for the development of nuclear weapons. Edward Teller, in particular, was following his main interest, the realization of the "H bomb", the fusion bomb which was announced as being several orders of magnitude more powerful than the fission bomb.

After three years teaching chemistry in Berkeley and having returned from Europe, Alder managed to get a permanent position at Livermore, no just as an advisor, as he had been up to that time. At first he accepted a part time job, dividing his time between Livermore and Berkeley. Then, since he was very much attracted by the idea of research work on computer, Alder gave up his position at the University of California and decided to spend all his time at Livermore.

However, within the Livermore group, Alder took up a unique position, far from the immediate problems of building new bombs: «Probably, I was the only one working with state equations during the first years in which weapons were being developed at Livermore», he recalls. That was a pioneering period: everybody did everything. Therefore, the young PhD also worked on the most experimental stages of development of the new centre. However, the main problem for Berni was still theoretical. The problem was still to ascertain whether a hard-sphere system could undergo a phase transition.

This question, as we have seen, had also been posed at Los Alamos, where no answer had been found. Alder, convinced that such a phenomenon was real, started to build up the idea that the reason for the difficulty was intrinsic to the Monte Carlo method, and that therefore a different method had to be developed. That is, a different way to use the computer had to be found for the simulation of many-body systems. At Livermore, he developed this idea and found a good partner to work on its implementation: Tom Wainwright, a 28-year old physicist, whose office was on the floor just below Alder's.

Thomas Everett Wainwright was born on September 22, 1927. He spent his early childhood in Southgate, a Los Angeles county, and later, in 1932, moved with his family to Montana where he spent his adolescence and attended high school. As soon as he was 18 years old, in 1945, he joined the US Army Air Corps and was sent to the island of Guam in the middle of the Pacific ocean, as a cryptography specialist. His engagement did not last long. The war was at its end, and the armies demobilized. Thus Tom could go back to Montana to study. In short, he obtained his bachelor degree in physics and engineering at the Montana State College and, in 1954, a PhD in physics at the University of Notre Dame, Indiana, presenting a dissertation on the electronic band structure of metals. He was brilliant, and was immediately offered a position at the Livermore National Laboratory. Here he met Berni Alder.

Tom Wainwright (1927–2007) (Oxford University Press)

«He was a modest and unassuming person, easy to interact with, but what particularly attracted me was his intelligence and inquisitiveness», Alder recalls.[19] The two physicists met at Livermore in 1955, just after Tom arrived from Indiana and Berni from Europe. Alder thought that, in order to solve his problems with the phase transition of a hard-sphere system in an original manner, he should try to go beyond the Monte Carlo method, which seemed useless for such a challenge. He was proven wrong on this point, because Bill Wood and his collaborators soon showed just the opposite, as we shall see later on. But this mistaken verdict was in the end a blessing, since it led Alder to explore a completely new road, one that eventually led him to success. The main reason why the Monte Carlo could not see phase transitions, Alder explained to Wainwright, was that it is a very effective method for calculating the equilibrium properties of a system. But this is a static condition. On the other hand, a phase transition is not an equilibrium condition. By definition, it is a dynamic process. Therefore, you need a new dynamic method, in order to calculate the way the

---

[19]B.J. Alder, *In Memoriam: Thomas E. Wainwright*, Progress of Theoretical Physics Supplement No. 178, 2009, pp. 1–4.

properties of a hard-sphere system undergoing a phase transition will change with time. Alder already had in mind an idea of what that method could be. Moreover, at Livermore, he would finally be able to meet the practical condition indispensable for realizing the project. Indeed, the centre established by Edward Teller had a computer which had the kind of processing power needed to solve this problem, so dear to Berni. Nobody else in the world at the time owned a more powerful computer, which was rather expensive. Therefore, the Livermore National Laboratory was keen to show that (so much) taxpayers' money had been spent on the right thing, and that the computer was working at full capacity. That is why they also offered processing time to groups doing research that was not directly related to the realization of nuclear weapons.

This powerful computer was therefore accessible and that was why Alder asked his young colleague, Tom Wainwright, to share the adventure. Tom accepted with enthusiasm and competence—better still, creativity. With his approach as a physicist and his experience of computers, Wainwright was indeed capable of transforming Alder's ideas, through a rigorous formalization, into a mathematical problem that could be solved. The idea was to find a method—different from Monte Carlo—in order to solve the problem of the phase transition in a hard-sphere system, and also, in general, to check the possibility of deriving, by computer simulation, the physical properties of time-dependent, i.e., dynamic systems, and hence study transport processes.

Instead of performing averages over random configuration samples, the idea was to follow the actual time evolution of a given initial distribution. Strictly speaking, this meant going back to Boltzmann's original approach. It was very probably a more expensive procedure in terms of computer time, but beside evaluating equilibrium properties, it would be able to study non-equilibrium phenomena by direct calculation of the time evolution of a given initial distribution of molecules. That sounded at the time like a prohibitive task for systems of molecules interacting via some kind of potential, but became feasible in a relatively simple way for a collection of hard spheres interacting only via elastic collisions. And, in its straightforward formulation, the method did not suffer from the uncertain probabilistic foundations that still affected the Monte Carlo approach.

Thus, Alder and Wainwright started using the Livermore computer to derive the equilibrium and non-equilibrium properties of the hard-sphere system, including the velocity autocorrelation function and Boltzmann's $H$ function, which are fundamental to understand the system dynamics. Thus a partnership was created, and it led to the new method of computer simulation which Alder had conjectured: molecular dynamics, a method able to simulate the evolution in time of a many-body system.

The idea is simple. When a dynamical system is made up of a relatively large number of interacting elements, its properties cannot be calculated with an "analytic method", i.e., exactly, by solving so many complex equations. This is expressed by saying that a dynamic system cannot be integrated with an explicit analytic expression: this is the most typical situation. That is why scientists use "numerical methods", i.e., they look for algorithms capable of providing accurate approximations. "Numerical methods" as a general rule require a great deal of processing power. They could never have been developed without computers.

However, now that they had an electronic computer with adequate processing power, Alder and Wainwright could exploit the power of the theory and use "numerical methods" to study the dynamics of a system made up of "many" hard spheres. In other words, the two scientists proposed to follow the evolution in time of this system by reconstructing the actual trajectories of each hard sphere in the phase space, calculating the positions and velocities of the hard spheres in time. Trajectories, which cannot be calculated analytically, can be computed numerically, that is by solving—step by step—the classical equations of motion, taking into consideration the collisions among the spheres.

In practice, one can start from an initial configuration of the system and follow the inertial trajectories of the spheres, up to the first collision between two of them. At this point, one calculates the velocities of the two spheres after the collision. The system is then allowed to evolve up to the next collision. Once again, one calculates the velocities after the new collision and lets the system evolve. The operation is repeated over and over again.

The electronic computer at Livermore worked hard. Pretty soon, the two scientists managed to get the first results on the time evolution of their system. The computing power of the machine was not yet high enough to enable them to accurately calculate properties such as the auto-correlation functions. However, Berni and Tom found a way to calculate the relaxation to equilibrium and to show that the $H$ function decays smoothly as a function of time. «I remember – Alder wrote on the occasion of the death of his friend, in 2007 – Tom trying to compare the behaviour of the $H$ function to the prediction of the Boltzmann equation, which involved a separate solution of this difficult integro-differential equation. It was an extremely complex task, but Tom succeeded, working day and night for several days». Finally, «the comparison worked well».

There was good agreement between the computer simulation and theory. Alder and Wainwright, with their new method of molecular dynamics, showed that the time required by a perturbed hard-sphere system to reach equilibrium is rather short, since it is just the time required for three to four collisions per particle. Thus they proved that it is possible to simulate the behaviour of the Boltzmann $H$ function, and check that it evolves almost monotonically toward equilibrium. Once the system has reached equilibrium, it is finally possible to calculate its thermodynamic properties. This meant that it was possible to accurately simulate the evolution of a many-body system. The only limitations on the accuracy of the results were the computer power and the exactness of the underlying physical theory.

The results obtained by Alder and Wainwright were so interesting that Charles Kittel, a prominent physicist working at Berkeley, included them immediately in the preface of his book on statistical mechanics. The results obtained with the new method—molecular dynamics—were first announced at the above-mentioned international meeting in Brussels in 1956. Their presentation generated «great excitement within the community of specialists in statistical mechanics».

Even John Kirkwood, who had abandoned the doubts he had first had about the solidity of the Monte Carlo method after the first papers published by Bill Wood at Los Alamos, now showed his enthusiasm for the new approach suggested by Alder and Wainwright, even before the presentation. The American scientists who had been invited by Ilya Prigogine left California for Brussels on a military aircraft. Everyone was on board that plane: Tom, Berni, and his former teacher, John Kirkwood. Alder took advantage of the situation to explain his results in detail. These were indeed the results he would present at the Brussels meeting. This time, Kirkwood did not have objections, and he did not start talking of his own integral equations. Instead, he realized at once that molecular dynamics was an insightful subject which he approached with enthusiasm. Berni Alder recalls spending hours and hours walking up and down the aisle of the aircraft on the way to Brussels, discussing the new method with his former teacher.

COLLOQUE INTERNATIONAL SUR LES PHÉNOMÈNES
DE TRANSPORT EN MÉCANIQUE STATISTIQUE
BRUXELLES, 27 AU 31 AOÛT 1956

Bruxelles Conference, August 27-31, 1956. First row, from the left, Maria Goeppert-Mayer (1906–1972), unknown, Ilya Prigogine (1917–2003), Lars Onsager (1903–1976); third row, behind Prigogine, Pierre-Gilles de Gennes (1932–2007) and, second to his right, Nico van Kampen (1921–2013). Berni Alder is in fourth row, between de Gennes and van Kampen; in the back, among others, John Kirkwood and Joel Lebowitz. (Photograph by J.R.P. Hersleven, courtesy AIP Emilio Segrè Visual Archives, Sengers collection)

However, while the proposal by Alder and Wainwright attracted the attention of the participants to the Brussels conference, and had already convinced Kirkwood, several other colleagues remained perplexed. Edward Teller, for example, stated that the method of molecular dynamics would never really compete with the Monte Carlo method, because of its extreme complexity.

All the same, Berni Alder and Tom Wainwright continued to work hard, spending thousands of hours in front of their computers. Having proved the reliability of their method, they started to refine it to get final results about the question of the phase transition. In a meeting in New Jersey in the month of January 1957, they showed the data they had obtained from systems of 256 and 500 hard spheres, and stated that they had seen a first-order phase transition from fluid to solid. They pointed out that this conclusion was different from the one obtained with the Monte Carlo method by Metropolis, by Rosenbluth, and even by Alder himself, in the work published in 1955 together with Frankel and Lewinson. They also added that the Monte Carlo results were being analyzed at the time by William Wood at Los Alamos.

The proceedings of the New Jersey conference were only published in 1963.[20] But soon after the meeting, in a very short lapse of time, Alder and Wainwright got convincing evidence that their hints were indeed proofs. In the month of August 1957, the *Journal of Chemical Physics* published the paper *Phase transition for a hard sphere system,* in which it was shown that the method of molecular dynamics solved the problem of phase transitions for both 32 and 108 hard-sphere systems.

In the same issue, the *Journal of Chemical Physics* published another paper, by William Wayne Wood and Jack David Jacobson, entitled *Preliminary results from a recalculation of the Monte Carlo equation of state of hard spheres.*[21] The two physicists at Los Alamos had used the Monte Carlo method and, by obtaining the same results as Alder and Wainwright, proved that the previous suspicions about the soundness of the Monte Carlo were not in fact justified.[22]

---

[20]B.J. Alder, T.E. Wainwright, *Investigation of the Many-Body Problem by Electronic Computers,* in J.K. Percus (ed.), *The Many-Body Problem. Proceedings of the Symposium on the Many-Body Problem held at Stevens Institute of Technology, Hoboken, New Jersey, January 28–29, 1957,* Interscience Publishers, New York, London 1963, pp. 511–522.

[21]W.W. Wood, J.D. Jacobson, *Preliminary Results from a Recalculation of the Monte Carlo Equation of State of Hard spheres,* Journal of Chemical Physics **27**, 1207, 1957.

[22]In fact, there was a great deal of collaboration between Alder and Wood. Indeed, before finishing the paper presented in the month of August 1957, Alder accepted Wood's invitation to Los Alamos and used the IBM 704, a new computer endowed with the highest processing power of the day. Alder worked for two months at Los Alamos, taking advantage of machine time—178 h out of a total of 704, representing 25% of the total available time, a remarkable proportion. Alder and Wood thus realized that their two different methods actually converged. In the end, the results they obtained appeared back to back in the same issue of the journal.

The IBM 704 computer. The 704 installed at Los Alamos was used by Alder and Wainwright in 1957 to perform the calculations needed to bring succesfully to completion their research on the hard sphere model (Lawrence Livermore National Laboratory)

And so there were now two possible ways—Monte Carlo and molecular dynamics—to carry out computer simulations of the statistical behaviour of many-body systems.

# Chapter 3
# The Growth of Molecular Dynamics

## 3.1 The Diffusion of Molecular Dynamics

Molecular dynamics was thus born between 1956 and 1957. Berni Alder and Tom Wainwright proved the existence of a phase transition from the fluid to the solid state in a hard-sphere system. It does not really matter that Wood and Jacobson obtained the same result with the Monte Carlo method. The important point is that the demonstration of the existence of a phase transition in the model hard-sphere system, using two different techniques, lent credibility to numerical methods.

However, Alder and Wainwright did not just show the potential of the processing power and speed of computers. They did much more than that. Indeed, they demonstrated for the first time that, thanks to the power of electronic processors, it was possible to study the evolution in time, that is, the dynamics, of a many-body system. A new subject matter was born, with its specific identity, but also a new way of doing science: computer simulation. In general terms, this meant not only the possibility of finding a numerical solution to whatever models had been invented, but also, in the narrower sense of the term, the ability to directly compute the consequences of the fundamental laws of physics.

«It was the centenary of Statistical Mechanics, and one of the directions of the second century was evident in the emerging numerical data on classical rigid sphere transitions», Jerome Percus stated shortly after this.[1] What he meant was that molecular simulation would open a new phase in the history of statistical mechanics.

However, this new approach was not immediately accepted by everyone with the same enthusiasm as Percus. Quite the opposite. At the end of the 50s, Percus, a well-known physicist and mathematician working at the Courant Institute of Mathematical Sciences in New York University, was rather the exception than the rule. Several other scientists—perhaps most of them—were sceptical in this respect. They were sceptical

---

[1]J.K. Percus (ed.), *The Many-Body Problem. Proceedings of the Symposium on the Many-Body Problem held at Stevens Institute of Technology, Hoboken, New Jersey, January 28–29, 1957*, Interscience Publishers, New York, London 1963, p. V.

© Springer Nature Switzerland AG 2020
G. Battimelli et al., *Computer Meets Theoretical Physics*, The Frontiers Collection,
https://doi.org/10.1007/978-3-030-39399-1_3

not so much about the results that had been obtained so far, as about the basic concept of computer simulation. As Berni Alder himself pointed out years later: «Of course, there has always been a resistance to use computers instead of theoretical methods. This feeling was very strong among the old timer physicists, to do things analytically instead of numerically».[2] However, this was not just a matter of principle. There were also practical motivations. «I think – Alder went on - that its slow acceptance was because people didn't have the computers to get into the game. Very few people had access to computers and that delayed things».

In sum, a significant number of scientists inside the community were behaving just like the fox in Aesop's fable, which, unable to reach the grapes, declared them to be unripe. In a word, it was sour grapes! Therefore, for the time being, Alder and Wainwright had to confirm, consolidate, and enlarge the results they had reached, in order to convince their community of peers. However, in order to do this, they needed more calculations. Berni recalled that, in those few months, he and Tom needed «a lot of statistics [...] that is why machine time at that point became "the" key factor».[3]

There was indeed, from the very beginning of their work, a different and possibly more crucial "key factor", one that had to do not with machine time, but with human manpower and skills. Neither Alder nor Wainwright, nor probably any one of the researchers working at Livermore and Los Alamos, knew how to translate their scientific ideas into the computer language. That is why the two labs hired "electronic programmers", people who knew machine language. Most of them were women, with a degree in maths.

These women played a crucial role. Without them, physics and chemistry would hardly have taken advantage of the new technological tool. In short, they constituted an integral part of the research teams. It is a shame that almost nobody ever recognized their role. Indeed, their names hardly appeared in published papers. Not even in the final acknowledgements. Many of these uncelebrated, yet invaluable contributors were simply forgotten. However, this did not happen with Alder and Wainwright. In their 1956 Brussels article, beside thanking Fernbach "for generously making computer time available", they also acknowledged the "extremely competent help" of Shirley Campbell on the 704, Leota Barr and Douglas Gardner on the Univac, and Ramon Moore and Donald Freeman in the machine solution of the Boltzmann equation. An otherwise unknown Mary Shephard appears together with Shirley Campbell as one of the coders that worked on the IBM 704 calculations reported in the three 1957 papers on the hard-sphere problem. The contribution given by Norman Hardy,

---

[2]Mac Kernan et al., *Interview with Berni Alder*, SIMU Newsletter no. 4, 2002, https://doi.org/10.13140/2.1.2562.7843, https://www.researchgate.net/publication/267979976_SIMU_Challenges_in_Molecular_Simulations_Bridging_the_Length-_and_Timescales_gap_Volume_4.

[3]G. Michael, *An Interview with Berni Alder. Stories of the Development of Large Scale Scientific Computing at Lawrence Livermore National Laboratory*, http://www.computer-history.info/Page1.dir/pages/Alder.html.

one of the first "computer scientists" at Livermore, was also acknowledged. However, the greatest collaborator of Alder and Wainwright was certainly Mary Ann Mansigh. She had studied mathematics, physics, and chemistry at the University of Minnesota, before answering a job ad in a magazine for a position as a software engineer at Livermore in 1955. She obtained the job even though "she was not a male". Three years later, Alder ensured her exclusive collaboration. Their partnership lasted twenty-eight years, during which Mary Ann displayed remarkable efficiency. Her task was not at all a minor one, namely, transforming Alder and Wainwright's physics ideas into a computer programme, and hence into machine language.

Mary Ann Mansigh in 1965

At the beginning, in particular, Mansigh's work was rather demanding, since each machine had its own language and machines changed all the time (the line used was IBM 700, 704, 709, etc.). So each time, Mary Ann had to translate what had been achieved with the previous machine into a new language. It was very hard work. Fortunately, between 1958 and 1959, everything changed. Indeed, Livermore became a place of experimentation for a new language, FORTRAN, invented by John Backus. This is a language that can be used on different machines. For Mary

Ann Mansigh this was a great relief. Although she still worked very hard, at least her work was not so boring, since she was no longer forever copying what she had been doing beforehand.

The development of molecular dynamics needed computing power, and hence a really huge amount of time on the computers available back then. This meant a far greater amount of time than was available—or at least accessible—at Livermore, given the fact that computers were meant to be used mainly for the official purpose of the laboratory, weapon development. However, as we all know, necessity is the mother of invention. The computers at the lab ran 24 hours a day. Mary Ann Mansigh made every effort in her programs to have no wasted time cycles and be very efficient. The capability of newer computers allowed there to be more than one program running on the computer and allowed the second one to use time when the original program had unused cycles. The programs were given priority numbers to decide the order of usage, and STEP (the early molecular dynamics program developed by Mansigh) had the lowest priority in each computer and only took up cycles not used by others. In spite of that limitation, Mansigh was able to have a background STEP program running on each of the computers, which allowed to pick up many unused minutes. Each one of the computer runs she made for Alder and Wainwright would require 40 hours of computer time to provide the correct amount of data. Over the years this resulted in the STEP program receiving very large amounts of computer time. Using this available time meant that Berni was the largest user of computer time at the lab for many years, even surpassing the time used by the weapons program.

All this work and calculation finally produced the desired fruits: a series of important scientific results, that were gradually diffused within the community. Moreover, in a manner of speaking, they also produced physics through pictures. This was probably an unforeseen side-effect: indeed, on the computer display used by Alder and Wainwright, unprecedented pictures appeared, which allowed them to view the hard-sphere motion and its differences in the solid and fluid state. Those pictures were so beautiful and significant that they stirred up surprise among all those who got to see them. In short, *Scientific American*, the world's most important journal for science communication, asked Alder and Wainwright to write an article addressed to the general public, accompanied by these pictures, which would speak for themselves. It was a great success. In less than two years, while still at the pioneering stage, molecular dynamics had already become a topic of general interest.

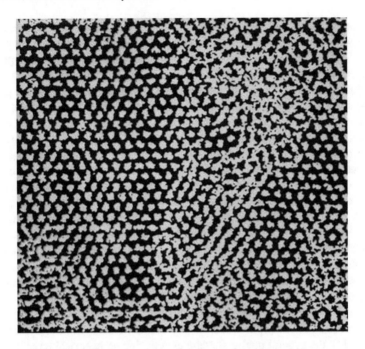

"This image of a two dimensional system shows the movement of the center of the particles and how starting from a regular pattern they slowly begin to melt, going from a solid state to a liquid. From the various prints on the desk we could see how larger particles remain in their structured solid positions, smaller particles move slowly over time and leave their original locations as they go to a liquid state, and the smallest particles representing a gas move around much more freely." (from M.A. Mansigh, The Early Years of Molecular Dynamics and Computers at UCRL, LRL, LLL, and LLNL, in Advances in the Computational Sciences. Symposium in Honor of Dr Berni Alder's 90th Birthday, Lawrence Livermore National Laboratory, 2015)

However, it is no easy matter to write a paper for the general public on a subject requiring specialist knowledge of both maths and physics. Indeed, the article by Alder and Wainwright was considered not really suitable by the director of the journal, Dennis Flanagan, who changed a few parts of the text to make it easier to read. But even Flanagan, experienced editor of the most authoritative journal for physics communication (he had been the director of *Scientific American* since 1947, and kept this role for a total of 37 years, until 1984), had difficulties "translating" from this highly technical context into common language. In sum, Flanagan had in turn to be corrected. The inaccuracies he had introduced while trying to simplify the paper were too many and often quite serious. «I remember Tom and I spending 8 hours on the telephone from the west coast to the east coast talking to the editor to get it straightened out».[4]

---

[4]B.J. Alder, *In Memoriam: Thomas E. Wainwright. September 22, 1927–November 27, 2007*, Progress of Theoretical Physics Supplement 178, 1, 2009.

Eventually, the paper was published.[5] Its title was *Molecular Motion,* and the abstract was apparently rather modest: «One of the aims of molecular physics is to account for the bulk properties of matter in terms of the behaviour of its particles. High-speed computers are helping physicists realize this goal». The authors went on with a qualitative analysis of both the Monte Carlo method and molecular dynamics, while quoting the results obtained so far and discussing the current limits of the new approaches. They then discussed prospects for the future, with a certain courage, since their statement was perhaps rather presumptuous: «it is not too much to expect that the information obtained by means of computing machines may play a role analogous to that of laboratory experiments in the development of the theory».

Their computer thus took the form of a virtual lab. Indeed, it would soon become clear that, in the balance between "sensible experiences" and "certain demonstrations", simulation would come more and more to constitute a strictly theoretical activity. However, at the time, Alder and Wainwright wanted to emphasize the fact that, thanks to computers, a new kind of knowledge could be produced. Be it a virtual experiment or a theoretical elaboration, new perspectives were certainly being opened: «When we have built up a sufficiently large body of numerical computations, we may be able to discern generalizations that are not apparent to us now».

The eight hours on the phone from west to east were not wasted. Indeed, *Scientific American* not only published the paper with which molecular dynamics was introduced to the general public, thereby announcing an historical breakthrough in the way science is practised; it also reproduced in the text one of the most significant pictures generated on the computer by Alder and Wainwright. That picture had already been presented at a meeting in Varenna in 1957, and was destined to appear in several textbooks soon afterwards.

In the *Scientific American* article, and later on in textbooks, the pictures were static, of course. However, Alder and Wainwright had created a dynamic system, which could visualize the particle motions. In other words, they produced a video. «In order to demonstrate the phase changes more vividly», they wrote in their paper, «a display system similar to a television picture-tube was hooked up to the computer. Each particle in the system was then represented by a dot on the face of the tube. By focusing a camera on the screen and leaving its shutter open, it was possible to record the trajectories of the moving dots on film. The photographs [...] show the imaginary material in both the crystalline and the fluid phases».

The idea was persuasive. Molecular dynamics could be visualized. One could "see" on a computer how microscopic many-body systems behave. This was not only important when communicating the new approach to science to the general public. It was also decisive in the peer-to-peer communication of molecular dynamics, among experienced colleagues. The display thus became a relevant feature of both the communication and the practice of molecular dynamics. In any case, as Berni Alder stated, this new feature favoured the acceptance by the scientific community of the new way to do research.

---

[5]B.J. Alder, T.E. Wainwright, *Molecular Motion*, Scientific American **201**, 131, 1959.

The use of computer-generated pictures of particle motions was also successful in school and university classrooms. Thanks to the possibility of "seeing" the particles move on a computer, the concepts and practice of molecular dynamics began to enjoy a remarkable diffusion in both school and university courses and textbooks.

Molecular dynamics in itself aroused the attention of both experts and the general public. However, the diffusion of new ideas was also helped by the general climate in the USA after the "Sputnik shock". In short, this is what happened. In 1957, the space era had started. The first man-made object was sent into orbit around the Earth. The first artificial satellite was launched by the Soviet Union. In the United States, this enterprise was considered a real slap in the face. Indeed, the mass media used the expression "Sputnik shock". How was that possible?

Many Americans—goaded on by the press, as well as by the statements of several Congressmen—began to denounce the "missile gap", i.e., the technological inferiority of the United States in the field of missiles, which clearly had great military importance. Sputnik showed for the first time that the USSR could reach US territory in a few minutes, with missiles carrying nuclear warheads. There was therefore a problem of national security, as well as the matter of prestige. At least, this was the prevailing rhetoric at the time.

It was not strictly true. Notwithstanding Sputnik, the United States still had a remarkable superiority in both science and technology in comparison with the Soviet Union. This was also true in the field of military science and technology. But the fact is that, in order to make up for the (alleged) technological gap, Washington elaborated and supported, among other things, a strategy of scientific education, at all levels, from primary schools up to universities. Thus, new methods for teaching physics were introduced in high schools, thanks to projects envisaged in the best universities, such as Harvard Project Physics, or the project proposed by the Physical Science Study Committee (PSSC), a group of experts brought together at the Massachusetts Institute of Technology (MIT) in Boston.

On the other hand, college students were targeted by the Feynman Lectures on Physics, put together at Caltech, Pasadena. These would be published in several volumes in 1964. In the same way, in Berkeley, a committee established in 1961 and chaired by Charles Kittel promoted the Berkeley Physics Course, a series in five volumes that was published in 1965. The fifth volume, entitled *Statistical Physics*, was written by Frederick Reif and carried on its cover a series of computer-generated pictures which showed the evolution in time towards the final state of equilibrium of a system of colliding particles. The first page of the volume was devoted to "Fluctuations in Equilibrium, Irreversibility and Approach to Equilibrium", a really unusual topic for a university textbook. However, the change was there, and the text was supplied with a series of pictures illustrating the basic concepts, as well as with the explicit acknowledgement: «All the computer-generated pictures were produced thanks to the generous collaboration of Doctor B.J. Alder of the Lawrence Radiation Laboratory at Livermore».[6]

---

[6]F. Reif, *Statistical Physics. Berkeley Physics Course*, vol. 5, McGraw-Hill, New York, 1965, p. 7.

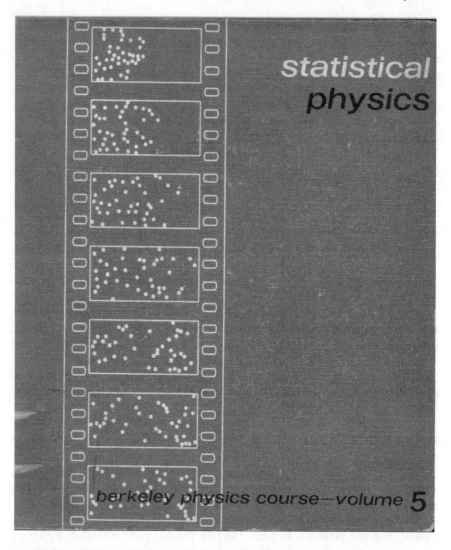

The front cover of F. Reif's *Statistical Physics*, fifth volume of the "Berkeley Physics Course" series, reproducing the images, computer generated by Alder and Wainwright, showing the irreversible evolution toward equilibrium of a system of colliding spheres (Mc Graw Hill)

## 3.2  New Computers, New Perspectives

However, this was not only a matter of public outreach and teaching. In that same year, 1959, in which the paper with pictures was published in *Scientific American*,

Alder and Wainwright sent a paper to *The Journal of Chemical Physics* with the title *Studies in Molecular Dynamics. I. General Method.*[7]

The paper presented a general method for studying molecular dynamics. Moreover, it proposed computer-generated pictures, an unprecedented feature for a scientific journal. The article listed the results obtained up to that time and indicated possible problems to be tackled in the future. It also announced that further papers would follow, in which those problems would be addressed and solved. And this promise would be kept: over the next twenty years, 18 articles would be published, all of them signed by Alder, together with different collaborators. These papers constitute a true series of *Studies in Molecular Dynamics*.

The main focus of the paper published by Berni and Tom in 1959 was "the bases of the algorithm" used to simulate the evolution of a system of classical interacting particles. This paper was of a general character. Alder and Wainwright remind the reader that the equations describing the behaviour of many-body systems do not usually admit analytic solutions. However, the difficulties are strictly of a mathematical, rather than a conceptual, character. Therefore, a computer with a sufficient computing speed seems to be the ideal tool in order to tackle them. For sufficiently dilute systems, the equations might be simple enough to handle, since in this case the behaviour of the system «can be conceived of as a succession of essentially unrelated binary interactions». The problem gets complicated in the case of non-dilute systems, when the range of action of the intermolecular forces is not small in comparison with the average distance between particles, and one cannot therefore simply consider binary interactions, but must take into account the interaction of each particle with all surrounding particles. The complication coming from the need to take into account the intermolecular interaction for dense systems does not exist in the case of the hard-sphere model on which Alder and Wainwright had worked until then. In that case, one always dealt with pair interactions, namely binary collisions among the model spheres. In order to solve that very problem, the two physicists developed and discussed in detail, in the introduction of their method, an algorithm which «is still very much in use nowadays», as Alder proudly remembered in 2009, on the occasion of Wainwright's memorial (he died in 2007), fifty years later. Needless to say, this happens only with really significant algorithms.

Almost all the physical and mathematical prerequisites for the development of molecular dynamics were already there. Moreover, in comparison with two years earlier, when the number of hard spheres of the system under investigation reached at most a hundred, processing power had risen significantly. Alder and Wainwright reaffirmed that the limiting factor in molecular dynamics seemed to be processing power alone. This problem could only be solved in two ways, either by waiting for more advanced computers, or by looking for "machine time", i.e., the possibility of using the existing computers for longer times.

Actually, Alder and Wainwright overlooked an important point. Insufficient computing power was not the only limiting factor. In fact, up to that time scientists had

---

[7]B.J. Alder, T.E. Wainwright, *Studies in Molecular Dynamics. I. General Method*, Journal of Chemical Physics **31**, 459, 1959.

only studied the properties of a highly unrealistic model system, made up of hard spheres interacting solely through direct collisions. A simulation of a more realistic model, representative of a real physical system, would require not only the as yet unavailable computing power, but also (and perhaps mainly) algorithms capable of getting the computer to complete the required calculations under the available conditions. In short, what was needed was both greater computing power and more theory. From a qualitative point of view, however, the problem was well defined in its general terms.

Meanwhile, computer technology was developing fast. Computing power and memory were growing all the time. In the article published in *The Journal of Chemical Physics* in 1959, Alder and Wainwright referred to the boost received by passing from Univac (the first computer they had used) to IBM 704. Livermore was soon to host IBM 7090, a truly advanced version of IBM 704, since it used transistors instead of valves. In 1960, they got LARC (Livermore Advanced Research Computer), the first purpose-built supercomputer realized by Remington Rand for the needs of the Livermore Laboratories. And even this did not hold primacy for long, since one year later it was overtaken by another IBM model, the 7030.

A picture taken in 1959 on the last day of operation of the Univac computer at Livermore. The computer had been used in 1956–57 for the early molecular dynamics simulations (Lawrence Livermore National Laboratory)

At this point, there were powerful enough computers at Livermore. However, the lab's priority was not the molecular dynamics of hard spheres, but rather calculations for the realization of more and more powerful and sophisticated atomic bombs. In other words, the cutting-edge computers at Livermore Laboratory were almost always busy, and rarely available to those who were not engaged in nuclear weapons.

It did not much help that Berni Alder had a good personal relationship with Sidney Fernbach, the director of the computation department at the Livermore Laboratory and a disciple of Robert Oppenheimer, who had been recruited into Teller's group in 1952. A quick career had brought him to the head of the computation department. In this role, Fernbach was interested in having the computer work full time. Thus he willingly allowed Alder and Wainwright the little remaining time on the machine, while restraining those who were committed to military research and objected that time was being wasted on research that was not strategic for the lab. Alder and Wainwright thanked him, and tried to achieve the greatest possible efficiency, so as not to lose one second of the precious, coveted time available to them, often just a few minutes.

However, neither machine time, nor programmers were sufficient at Livermore. They had to look for precious machine time elsewhere. In 1957, therefore, Alder left Livermore to seek machine time at Los Alamos. This was no accident. It was still the case that there were only a few powerful computers, mostly in military research centres.

However, not all of them were in such centres. In fact, our two pioneers of molecular dynamics were helped in their search for machine time by several people. Among others, there was Norman Hardy, who developed a highly efficient algorithm for making better use of machine time. In short, this consisted in updating, during calculations, a list of particles next to the ones under examination, and thus considering only collisions between nearby particles, so that it was no longer necessary to take into account more remote particles, and precious time could be saved. Hardy's scientific contribution ushered in a whole series of intelligent algorithms.

Among others, people like George Michael gave them a huge amount of machine time. Michael was a climatologist, a former colleague at Livermore, who was working in the Pacific islands. He had a powerful computer at his disposal, and he also had a lot of machine time available. The task of calculating the weather forecast in that area used Michael's machine for barely 15 min a day. Therefore, his computer was free for the remaining 23 hours and 45 min, i.e., practically all the time. Michael allowed Alder and Wainwright to use that time with no scruples whatsoever. Such generosity was really useful to molecular dynamics research and the two scientists did not miss this opportunity.

There is more to it. As we mentioned earlier, in the final part of a paper published in 1959, Alder and Wainwright reminded the reader that simulations could be picked up by a cathode ray tube connected to the computer and, with a simple algorithm, pictures could be obtained from numerical calculations. The molecular dynamics of a computer-simulated system could now be visualized. This possibility affected not only the way scientists worked, but also influenced appreciation of such work by the general public. Mary Ann Mansigh created a program that allowed to produce a movie about irreversibility with purely didactic aims. The video further supported the pictures that Alder was offering to illustrate the fifth volume of the "Berkeley Physics" series, devoted to statistical physics. He also had a new idea, which was particularly innovative at the time: why not sell the book and movie together? «I don't think it was a commercial success, but it was used a lot. Many fashionable physics professors have told me they have used that demonstration to teach the concept of irreversibility», Alder recalls.[8]

Berni Alder, Mary Ann Mansigh and Tom Wainwright looking at the computer generated pictures showing the phase transition in a system of hard disks, 1962 (courtesy AIP Emilio Segrè Visual Archives)

By this time, the computational sciences were growing, hand in hand with the development of electronic computers—the first transistor computer, ELEA 9003, was built by Olivetti in Ivrea in 1957. In particular, there was a growing interest in the subject of radiation damage in metallurgy. In 1960, a group of researchers at the Brookhaven National Laboratory in Upton, in the State of New York, published a paper on the *Dynamics of Radiation Damage*.[9] The group effectively simulated on a computer the damage produced by radiation upon a model of solid copper, showing that some nuclei are ejected from their lattice sites due to their impact with radiation. They thus tested the consequences of these ejections, which may be of two different

---

[8] Alder interview cited in note 3, p. 4.

[9] J.B. Gibson, A.N. Goland, M. Milgram, and G.H. Vineyard, *Dynamics of Radiation Damage*, Physical Review **120**, 1229, 1960.

kinds: they may introduce either interstitial atoms, i.e., extra atoms between the lattice atoms, or create voids inside the lattice. The latter are called "lattice vacancies".

In Livermore, the LARC (Livermore Advanced Research Computer) became operational in June 1960. It was the first supercomputer ever made. The computing power available to Alder and Wainwright was thus significantly increased. Their desire to spread their new method also increased. In 1963, Alder inaugurated a series of books, *Methods of Computational Physics*, entirely devoted to computer simulation. The idea was to publish all the material gathered in Los Alamos and the Lawrence Livermore Laboratory as soon as the constraints of military secrecy were lifted. The result of this work was a series in twenty volumes, each devoted to a specific area of study, from statistical mechanics to quantum physics, and from fluid dynamics to solid-state physics. «This was the first, extremely important project, which managed to involve more people in our research field», Berni Alder recalls.

In that same year Alder helped Edward Teller set up the Department of Applied Science at the University of California at Davis. Its aim was to train young people for his "new physics", i.e., computer simulation physics.

Only a few years had gone by since its first formulation. Molecular dynamics was still in its pioneering stage. In this period, people were trying to test the validity of the fundamental hypotheses of statistical mechanics, such as Ludwig Boltzmann's "H theorem". They were also studying phase transitions, developing more and more sophisticated algorithms to integrate the equations of motion, and formulating the state equation for model substances, such as hard-sphere systems, as a model for simple liquids. However, something important was still missing, viz., simulations of the behaviour of real systems.

Of course, it was a surprise that a phase transition could occur within a hard-sphere system where spheres interacted only through elastic collisions, and even more so that it could be "viewed" on a computer, as Alder and Wainwright had shown. Nonetheless, the idea that one could use all the tools of statistical mechanics to simulate the realistic dynamic behaviour of both atoms and molecules on a computer, bringing in other kinds of interactions beside simple elastic collisions, took time to get established. In fact, it took eight more years and the appearance on the scene of a researcher of Indian origin, Aneesur Rahman.

Thanks to Raman and his paper, *Correlation in the Motion of Atoms in Liquid Argon*, published in *Physical Review* in 1964,[10] the development of the molecular dynamics approach accelerated dramatically.

---

[10]A. Rahman, *Correlation in the Motion of Atoms in Liquid Argon*, Physical Review **136**, A405, 1964.

## 3.3    Aneesur Rahman and Molecular Dynamics for Real Systems

With his 1964 paper, Aneesur Rahman laid a cornerstone in the history of computer simulation and became one of the founding fathers of molecular dynamics. Indeed, his paper was an inspiration to many, since, as Loup Verlet wrote, it «opened a new field of research, and originated a lineage of scientists to which we belong».[11] In short, this meant an unprecedented acceleration in the study of molecular dynamics. This new field concerned the computer simulation of real systems. The next generation of scientists would thus use computer simulations, not only for the dynamics of abstract hard spheres, but also for real systems of both atoms and molecules.

Aneesur Rahman (Anees to his friends), the new star of molecular dynamics, was born in 1927 in Hyderabad, a city in today's Federal State of Telangana, in the heart of India. He belonged to a cultured and influential Islamic family. His mother, Aisha, had quite powerful relations in the city. His father, Professor Habibur Rahman, was an outstanding figure in the Nawayath, a Muslim community typical of southern India. Professor Habibur was at the same time prominent and generous, and was considered as such in his environment, since he donated his own property for the establishment of the Urdu Arts College, as well as for the Urdu Hall, so as to encourage the cultural development of young people within his community.[12]

The influence and reputation of the Rahman family in Hyderabad were reinforced when, between 1965 and 1972, Anees' brother, Fazlur, became vice-Chancellor of Aligarh Muslim University, a public university which was defined as an institute of national importance.

In this environment and belonging to this family, Anees received a good education, which allowed him to obtain a diploma in mathematics at the high school in his home city in 1946. He then moved to England, where he took the "tripos" in mathematics and physics at the University of Cambridge in 1948 and 1949.

Once he had got his degree, he left the United Kingdom (he felt ill at ease in the old-fashioned imperialist atmosphere that still reigned there) and moved to continental Europe, to the University of Leuven in Belgium, where he worked with Charles Mannenback, a chemical physicist. In 1951, he met Yueh-Erh Li, a Chinese woman attending the medical school there, and fell in love with her. In 1953, Anees Rahman obtained a PhD in Physics at the age of 26. He went back to India to teach physics at Osmania University in Hyderabad, and continued his research at the Tata Insti-

---

[11]L. Verlet, *The Origins of Molecular Dynamics*, in J.P. Hansen, G. Ciccotti, H.J.C. Berendsen (eds.), *In Memoriam. Aneesur Rahman. 1927–1987*, CECAM, Orsay 1987.

[12]K. Sen, S. Sastry, *Aneesur Rahman. A Pioneer in Computational Physics*, Resonance, August 2014, pp. 671–683.

tute for Fundamental Research (TIFR) in Bombay, a government research centre established in 1945, under the authority of the Department of Atomic Energy. During the following four years, Rahman published several papers on the structure of diatomic molecules. Meanwhile, in 1956, he obtained a three-year National Postdoctoral Fellowship, which allowed him to work as a full-time researcher at the Physics Department of Osmania University.

In this period, he developed an interest in the structure of a rather important and also chemically peculiar triatomic molecule: water. For his research, Anees only had access to a mechanical calculating machine, FACIT. Working conditions in India were very different from those of his colleagues in the US and Europe. Rahman realized this, and always stayed in touch with Charles Mannenback's laboratory, determined to pursue cutting-edge science.

On November 3, 1956, when he was almost 30 years of age, Anees went back briefly to Belgium to marry his Chinese fiancé, Yueh-Erh Li, who had in the meantime obtained her degree in medecine. He then immediately went back to India. Just one year later, on December 17, the young couple had a baby girl, Aneesa.

In 1958, Anees moved with his family to Bombay, to work at the Tata Institute for Fundamental Research. This institute was considered at the time to be the best in India. TIFR was involved in fundamental research: physics, chemistry, biology, mathematics, and computer science. Rahman stayed two years in Bombay. Then he went back to Belgium, since he had been hired by IBM, which had recently inaugurated its research centre in Brussels. By now, computers—the most advanced computers—had become the tools of his trade. Anees also taught at Digital Arts & Entertainment (DAE), a computer science school in Howest University College, in Western Flanders. Meanwhile his wife, Dr. Yueh-Erh Li, was able to work in Leuven, at the laboratory of Christian de Duve, a first-class researcher, who obtained the Nobel Prize for Medicine in 1974.

Once again, however, the Rahmans did not stay there long. Indeed, in the month of March 1960, Aneesur left Europe and moved to the United States, accepting an invitation from the Solid State Division of the Argonne National Laboratory. Rahman welcomed this new job, and his second change of continent, also because his wife Yueh-Erh Li obtained a position at the Biomedical Division of a large research centre, a branch of the Metallurgical Laboratory where Enrico Fermi had worked, established in Illinois, 40 kilometres from Chicago, in 1946.

Anees Rahman (1927–1987) in a picture taken in 1967
(Springer)

Rahman stayed at the Argonne National Laboratory for 25 years. At first, he was involved in spectroscopy, and tried to elaborate a new theory. Later on, he became a pioneer of computer simulations of physical systems, and one of the founding fathers of molecular dynamics.

We thus arrive in 1964, the year he published his paper in the *Physical Review*. This is how it starts: «A system of 864 particles interacting with a Lennard-Jones potential and obeying classical equations of motion has been studied on a digital computer (CDC 3600) to simulate molecular dynamics in liquid argon at 94.4 K and a density of 1.374 g cm$^{-3}$».

The Lennard-Jones potential is calculated empirically, and is largely used to describe molecular interactions. The fundamental idea is that, when two atoms (or two molecules) get very close to each other, the electronic clouds (strictly speaking, the electronic densities) surrounding the two nuclei superimpose, thus generating remarkable repulsive forces which have a very short range, but grow quickly as the two atoms get closer. To these, London's dispersive (attractive) forces are added: these are instantaneous dipole–induced dipole interactions between non-polar molecules. They act at relatively short distances, distinctly longer, however, than the repulsive forces.

There is no theoretical equation describing the repulsive forces. That is why one must use empirical potentials. The potential derived by the English physicist John Lennard-Jones is the one most widely used.

This is the novelty: Anees Rahman did not simulate the molecular dynamics of an extreme model, such as the hard-sphere system, on a computer, but focused rather on a realistic model of a physical system, made up of real (argon) atoms. Thus the Indian scientist inaugurated a crucial new research field which opened up molecular dynamics and computer simulation to the study of molecular systems.

The difference with Alder and Wainwright's model is clear. The only interaction between the hard spheres taken as a model by the two researchers at Livermore was the collision when they hit each other. On the other hand, Rahman's real atoms "felt" each other, both through the repulsive forces generated by the superposition of their respective electronic densities and through London's attractive forces.

Rahman solved the classical equations of motion of the argon atoms, confined in a given volume with a density and temperature typical of the liquid state, thus succeeding in calculating a series of structural and dynamical properties of the system, which he then compared with the values obtained experimentally, by observing a similar system of liquid argon. Simulation and experiment agreed.

The scientific community immediately acknowledged this result. In 1965, just one year after the publication of Rahman's paper, Robert Zwanzig, a well-known expert in statistical mechanics, wrote: «Rahman's calculations provide the most detailed "experimental" information currently available about dynamic processes in liquids».[13]

Zwanzig thereby affirmed that computer simulation provided "experimental" evidence, and as Kalidas Sen and Srikanth Sastry remarked, announced the twofold key role that computer simulations of molecular dynamics would play in the following decades, both in terms of almost experimental evidence, capable of providing detailed information about the behaviour of a system of atoms or molecules, and in terms of theoretical prediction. This was a genuine phase transition. Indeed, molecular dynamics is theory, but it is so powerful that it gets very near to experiment.

A peculiar route led Anees Rahman to study for the first time the molecular dynamics of a real system. Since he was in Argonne, in fact, Rahman had become interested in the theory of neutron scattering in condensed matter, in particular, matter in the liquid state. What Rahman wanted to discover was how neutrons were deflected when they hit liquid matter. His studies were very thorough. Very soon Rahman elaborated a theory of neutron scattering in liquids, thus placing himself at the cutting-edge in this field of study.

Up to this time, Rahman had not been particularly interested in statistical mechanics and knew very little about molecular dynamics. In his 1964 paper, he did not even

---

[13]Quoted in K. Sen, S. Sastry, *Aneesur Rahman. A Pioneer in Computational Physics*, Resonance, August 2014, pp. 671–683.

mention the papers by Alder and Wainwright. It seems likely that he had not even read them.

However, since he was not satisfied with his own theoretical approach to neutron scattering, he started searching for new models which would allow him to check its validity. This research steered him towards computer simulation of the dynamics of liquids, using procedures of numerical integration of the classical equations of motion of individual atoms. In other words, he independently reinvented molecular dynamics, applying it to a realistic model of a system of atoms. And he achieved that notwithstanding all the novelties involved.

Of course, he considered argon, among the least reactive of atoms, as similar as possible to hard spheres, whose interactions were well described, though nevertheless approximately, by the empirical Lennard-Jones potential. Unlike hard spheres, these are atoms endowed with electronic clouds which, as they get nearer, generate both repulsive and attractive forces, the ones that must be taken into account. And this is precisely what Rahman did.

His work was not only the first simulation of the molecular dynamics of a realistic model, no small matter in itself. It was also a study of the liquid state, a state which, at the beginning of the 60s, could still not be handled from a theoretical point of view, and had hardly been investigated from an experimental point of view. In sum, the physics of "condensed matter" was taking its first steps. The only exception at the time was harmonic crystals, for which a theory already existed.

Rahman's work was no small contribution to that development. In short, his work provided a preliminary but clear demonstration of what computer simulation could do: it could realize "computer experiments" in which detailed information about the system under study could be calculated and compared both with experimental results and with the predictions of analytical theories, where applicable. Rahman calculated the theory with such precision that he allowed a direct and detailed comparison with experimental data.

This was a real success. However, seven years had elapsed between the first paper published by Alder and Wainwright about the molecular dynamics of hard spheres and Rahman's paper on the molecular dynamics of real argon atoms. Why had it taken such a long time to go from the study of an ideal hard-sphere model to a system of real atoms? This is what Loup Verlet refers to as the «7-year gap»: the gap between two phases of a process which, a posteriori, seems almost to be taken for granted, natural, and immediate.

Science hardly ever follows linear or obvious paths. Its historical trajectories almost always have many causes. One explanation for the «7-year gap» was indicated by Alder and Wainwright in their 1957 paper: «Although it is feasible to deal with realistic potentials, it entails a considerable slowing down of the calculation and involves the problem of having to cope with repulsive collisions where the forces the particles experience change very rapidly». Therefore, it was not only a question of computing power. There was a physical problem that had to be faced, namely, the repulsive forces due to the electrons, which change quickly, depending on the distance between atoms. There was a mental barrier here: the idea that "many-body" problems don't have exact solutions.

Anees Rahman was the first scientist to face up to this physical problem and break through the mental barrier. Indeed his paper, published in 1964, was the starting point that «led to developments related to construction of potential models for other real systems and the use of molecular dynamics to explore more complicated systems, in particular molecular fluids».[14]

That computer simulation had become an invaluable tool for the physics of molecular clusters was shown in the paper *Structure of water. A Monte Carlo calculation* published by John A. Barker and Robert O. Watts, two Canadian researchers (who were actually born in Australia), working in the Department of Mathematics and Applied Physics at the University of Waterloo, Ontario (Canada). In this paper, they studied the best-known fluid and derived, using an adequate model of the potential, the energy, the specific heat, and the radial distribution function of water at a temperature of 25 °C. Their results agreed well with experimental data.[15]

This particular work used the Monte Carlo method rather than molecular dynamics. However, it was clear by then that computer simulation was raising more and more interest. The conditions were right for the informal establishment of a new scientific community, cutting across classical disciplines, since it involved both mathematicians and physicists, chemists, and biologists. Berni Alder gave this new community a further reference point by establishing a new scientific journal in 1966, the *Journal of Computational Physics*, which would cover all aspects of computation relating to physical problems, presenting all the most significant new techniques of numerical solution to problems in every area of physics. This journal was particularly attentive to methods crossing the borders between the various traditional fields of scientific knowledge, and was in itself interdisciplinary.

The *Journal of Computational Physics* complemented the book series *Methods of Computational Physics* which had been established four years earlier. Thus, by the mid-60s, a group of scientists started meeting at conferences and publishing papers in the same specialized journals and books in the same series. This was a clear sign that a new scientific community had been created, namely, the molecular simulation community.

## 3.4 Alder's Discovery and Rahman's Water

Meanwhile, there were more and more computer simulation studies of the molecular dynamics of fluids using the Lennard-Jones potential. Among the protagonists of this research was the Frenchman Loup Verlet, a disciple of Victor Weisskopf. In 1967, he published a paper with the title *Computer Experiments on Classical Fluids,* in which he introduced both the algorithm currently bearing his name, for the numerical

---

[14]R. Kapral, G. Ciccotti, *Molecular dynamics: an account of its evolution*, in C. Dykstra et al., *Theory and Applications of Computational Chemistry*, Elsevier, 2005, pp. 425–441.

[15]J.A. Barker, R.O. Watts, *Structure of water; A Monte Carlo calculation*, Chemical Physics Letters **3**, 144, 1969.

integration of the Newtonian equations of motion (Verlet integration), and the Verlet list, a list of neighbours of each individual particle which is used to speed up the calculations of the interactions among molecules.[16] We shall say more about Loup Verlet later on.

In that same year, Alder and Wainwright published a new paper about the "long-time tail" of the velocity auto-correlation function.[17] For several years, the two scientists had been fully exploiting the potential of the new generation of electronic computers, to improve and consolidate the preliminary results obtained in the study of dynamic and/or non-equilibrium processes by the mid-50s. In particular, they had long been studying the process through which fluid particles, after casual collisions with other particles, gradually lose memory of their initial conditions, obtaining results reinforced by a more and more robust statistics. The mathematical object embodying this piece of information is the velocity auto-correlation function. The widely recognized assumption at the time was that, thanks to the intrinsic randomness of the process, the velocity auto-correlation should decay exponentially in time. Contrary to this widespread expectation, from their simulations Alder and Wainwright started getting more and more compelling evidence that this was not the case. A particle in a fluid does not lose its initial velocity exponentially; rather, it maintains it for a long time, even after several collisions with other particles. It is as if, in a way, it keeps a recollection of its initial velocity.

Alder recalls: «We discovered something unusual. I mean everybody since Einstein said that the decay had to be exponential because one hoped that eventually information is forgotten so that after sufficiently long time processes should become Markovian, leading to an exponential decay of this function, and we found instead this tail... The question then was: "What is the origin of this?". The physical problem we face here is that a particle going initially in some direction prefers to continue in that direction after 100's of collisions; we could not go much more than a 100 collisions. How could that possibly be? Tom and I were talking about that every day, just trying to understand this qualitatively».[18]

This phenomenon was so new and unexpected that the two scientists waited for a long time before publishing their results. Indeed, they had to show indisputable evidence, and needed a physically sensible model to explain this behaviour, which went against any reasonable prediction.

---

[16]L. Verlet, *Computer "Experiments" on Classical Fluids. I. Thermodynamical Properties of Lennard-Jones Molecules*, Physical Review **159**, 98, 1967.

[17]B.J. Alder, T.E. Wainwright, *Velocity Autocorrelations for Hard Spheres*, Physical Review Letters **18**, 988, 1967.

[18]B.J. Alder, *Concluding Remarks: The Long-Time Tails Story*, in M. Mareschal, B.L. Holian (eds.), *Microscopic Simulations of Complex Hydrodynamic Phenomena*, Plenum Press, New York 1992, pp. 425–430.

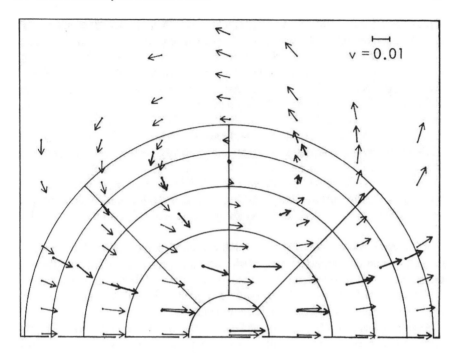

The velocity field (statistically averaged) around a central disk, clearly showing the coherent
vortex pattern. The figure convincingly indicates the good agreement between the velocity field
obtained from computer-simulated molecular dynamics (heavy arrows) and that given by the
hydrodynamic model (light arrows). The picture is taken from the 1970 paper by Alder and
Wainwright in Physical Review A. (Copyright 1970 American Physical Society)

An attempt at an explanation was finally put forward by Wainwright, using the
hypothesis that, even at the microscopic level, there was a mechanism, well-known
on a macroscopic scale in hydrodynamics, where it is known that in a fluid, under
certain conditions, coherent swirling structures can be created, which show a high
degree of correlation. Alder recalled: «We puzzled over that result for over 2 years.
We first did not believe the result and made sure the program gave the correct result.
After that, I remember going to Tom's office every day, trying to find an explanation.
Tom finally came up with the idea that it must be a hydrodynamic phenomenon since
nothing else could have that long a memory. Indeed, he, being much more familiar
with hydrodynamics than I, worked out the details. We confirmed the existence of the
vortex that caused the effect by Molecular Dynamics simulations and quantitatively
compared it to a separate hydrodynamic calculation. The result, nearly immediately
confirmed by graph theoretical considerations, revolutionized transport theory and
even more importantly totally destroyed the belief that hydrodynamics does not apply
close to the atomic scale».

Alder and Wainwright used this theoretical (hydrodynamic) explanation and presented their recent work in Kyoto, during a conference of the International Union
of Pure and Applied Physics (IUPAP). Their journey towards the old capital city of

Japan was rather eventful: Wainwright's luggage was lost. As a consequence, a rather amusing episode followed. Tom could not easily adapt to unexpected situations and did not accept either the idea of wearing the same clothes each day or of taking up Berni Alder's offer to lend him his own clothes. He wanted to buy new clothes. However, it was not so simple in Japan in the 60s to find clothes for a man taller than 180 centimetres. Alder and his wife solved this problem: they asked a couple of shop assistants, who could not speak English at all, to lay all the clothes they had upon a ladder so that Wainwright could choose the largest size of the available models.

So finally another cornerstone of Alder and Wainwright's work could be presented with sufficient elegance to the scientific community. Once again, they demonstrated the power of visualization as a crucial support: the published paper was completed by pictures, clearly and effectively showing the agreement between the data obtained by the simulation of molecular dynamics and those derived from a hydrodynamic model.

This paper also proved to be an inspiration, since on the one hand it stimulated the publication of a long series of articles seeking a rigorous theoretical basis for the phenomenon observed by Alder and Wainwright, while on the other hand, it demonstrated that hydrodynamics could be applied on the microscopic scale. This concept had been simply unthinkable before that.

One year later, in 1970, Alder and Wainwright wrote their eighth and last joint paper.[19] It concerned the calculation of transport coefficients within a fluid hard-sphere system. Their year-long partnership ended in 1971 when Wainwright became director of the Department of Theoretical Physics at Livermore Laboratory, and no longer had time for "his" molecular dynamics. He was compelled to take care of other matters.

Meanwhile, Anees Rahman came back on the scene. In collaboration with Frank Stillinger, he published a paper on the molecular dynamics of a realistic system,[20] made up not of single atoms, like argon, but rather of hard polyatomic molecules, viz., water, whose structure had started to interest him fifteen years earlier in India.

The compound $H_2O$ comprises three atoms, two atoms of hydrogen and one of oxygen. This implies a remarkable complication in comparison with monatomic argon. In short, Rahman and Stillinger simulated the evolution according to classical dynamics of a system comprising 216 water molecules at a temperature of 34.3 °C and studied in detail both its static structural properties and its dynamic behaviour. The comparison with experimental data was considered good by the two authors.

This was yet another ground-breaking paper. Once again, they managed to speed up the development of molecular dynamics. Its relevance cannot be confined solely to the technical data mentioned above, however significant. In fact, water is an extremely interesting molecule. It is very common on Earth and it can be found practically everywhere. Indeed, we find it in both seas and oceans, in rivers and lakes, in glaciers

[19] B.J. Alder, D.M. Gass, and T.E. Wainwright, *Studies in Molecular Dynamics. VIII. The Transport Coefficients for a Hard-Sphere Fluid*, Journal of Chemical Physics **53**, 3813, 1970.

[20] A. Rahman, F. Stillinger, *Molecular Dynamics Study of Liquid Water*, Journal of Chemical Physics **55**, 3336, 1971.

and deep underneath the Earth, but also in all geological, chemical, and—above all—biological systems. Water has unique properties. For example, it is a universal solvent. Furthermore, it presents a series of anomalies which are not all well understood by chemists, notwithstanding the fact that the $H_2O$ molecule has been intensively studied for over two centuries. Indeed, it has probably been studied more than any other molecule.

Between the end of the 60s and the beginning of the 70s, there has been an ongoing discussion among chemists about the microscopic structure of water in general and, in particular, about the structure of fluid water. Several complex models have been made, including "continuous", "mixed", "interstitial", and so on.

In simple terms, the problem lies almost entirely in the nature of the "hydrogen bond", which definitely plays a leading role in water. This is due to the fact that oxygen is a strongly electro-negative gas, and this means that it has a remarkable ability to attract electrons, partly removing them from the two hydrogen atoms in the water molecule. Thus the $H_2O$ molecule becomes polarized. In particular, the oxygen assumes a partially negative charge and the two hydrogens a partially positive charge.

The two hydrogens of one molecule may thus be attracted by the oxygen of another molecule, thus forming the "hydrogen bond", which is relatively stable. Moreover, the pyramid-shaped geometry (tetrahedral, according to chemists) allows each water molecule to take part in four "hydrogen bonds" and determine a series of anomalous properties of water, including a certain "structure" within the substance when it is in its liquid state. And so it was that the scientific debate among theoretical chemists about the "structure" of water remained alive throughout the 60's and 70's. The problem had not yet been solved.

The molecular dynamics study by Stillinger and Rahman proved that it was possible to reproduce quantitatively a wide range of properties of liquid water, from its structure to its thermodynamic properties and micro-dynamics. This was an unexpected and, to some extent, disruptive result. Alfons Geiger commented that no theoretical chemist could have foreseen it.[21] At least, not in the near future.

As so often happens in science, the partnership and, later on, the joint work of Rahman and Stillinger originated by chance. Indeed, the two scientists met in 1969 at a "Gordon conference". It was one of a series of prestigious conferences, which established themselves through their special format, that is, a limited number of invited participants, and free discussions during informal meetings behind closed doors. The conference in which Rahman and Stillinger met was devoted to "Chemistry and Physics of Liquids". The Indian scientist explained his pioneering work on the molecular dynamics of argon. On the other hand, Stillinger talked about his work on a model of water which he was developing with a colleague, Arieh Ben-Naim. Their discussion soon became a programme: computer simulation of the structure, dynamics, and thermodynamics of water under the most diverse conditions. This was a work programme which, in their minds, would probably go on for a long

---

[21] A. Geiger, *The Simulation of Water and Aqueous Solutions by Aneesur Rahman*, in: J.P. Hansen, G. Ciccotti e H.J.C. Berendsen, *In Memoriam. Aneesur Rahman. 1927–1987*, CECAM, Orsay, 1987, p. 21.

time, perhaps even for decades. Indeed, water plays a particular and crucial role in our living environment. Therefore, if we want to study the molecular dynamics of biomolecules, that is, molecules of biological interest, we must deal first of all with water and its peculiar behaviour.

The result of this key meeting between Stillinger and Rahman appeared in 1971 in the *Journal of Chemical Physics* under the title *Molecular Dynamics Study of Liquid Water*. This paper gave renewed impetus to the development of molecular dynamics.

A year earlier, in 1970, in the *Physical Review*, a paper had been published on the computer simulation of a system composed of liquid carbon monoxide (CO).[22] It was clear, even before Rahman and Stillinger's work, that molecular dynamics had entered a new phase: the systematic study of rather complex real systems, like those involving polyatomic molecules. However, as Alfons Geiger writes, the increased rate of development and the general acceptance of molecular dynamics for the study of systems in both chemistry and biology took place after Stillinger and Rahman had published *Molecular Dynamics Study of Liquid Water*.

---

[22]G.D. Harp, B.J. Berne, *Time-Correlation Functions, Memory Functions, and Molecular Dynamics*, Physical Review A **2**, 975, 1970.

# Chapter 4
# Molecular Simulation Arrives in Europe

## 4.1 France: Loup Verlet

«Immediately after World War II, the increase in power of computers provided a way to reconsider the question raised by the atomic hypothesis and addressed by Maxwell and Boltzmann in the 19th century: how to determine the macroscopic properties of gases and liquids, knowing that they are composed of large numbers of atoms which obey Newton's equations of motion, assuming that they interact via a given force law. This question became all the more pressing towards the end of the century, in light of growing evidence that atoms and molecules really exist. Until the end of World War II, this question remained essentially unsolved, except in the case of low density gases [...] The situation changed in the 50s due to the development of powerful computers. It became possible to follow, over a sufficiently long time span, the trajectories of several tens or even hundreds of particles behaving like billiard balls. A few years later it became feasible to tackle a more realistic model, containing about one thousand particles interacting via pairwise Lennard-Jones forces. Within a collective research effort, I suggested, among other procedures to speed up the calculations, a new algorithm for the integration of Newton's equation of motion of the dynamics of the molecules. Simple, stable, and accurate, this algorithm can be used to simulate the motion of very complex molecules».[1]

This brief description of the historical background of his pioneering work was proposed by the French physicist Loup Verlet. His studies introduced molecular dynamics into Europe for the first time and immediately made him famous. But in fact, Verlet's work was not a bolt from the blue. On the contrary, the French physicist was part of a more general climate which at the time was drawing the attention of many European scientists to the latest developments in computer simulation. Apart from Verlet and his research group, we may mention also Ian McDonald and Konrad Singer in Great Britain, and Kurt Binder in Germany among the main actors of molecular simulation.

---

[1]L. Verlet, *Chimères et Paradoxes*, Les Éditions du Cerf, Paris, 2007, p. 124 and pp. 174–175.

© Springer Nature Switzerland AG 2020
G. Battimelli et al., *Computer Meets Theoretical Physics*, The Frontiers Collection,
https://doi.org/10.1007/978-3-030-39399-1_4

As we shall see later on, these physicists made important contributions to the development of molecular simulation in Europe, and obviously also outside Europe. However, Loup Verlet also realized the novelties of simulation in both method and epistemology. Indeed, simulation opened up new perspectives to physics, but also to other research fields, from chemistry to biology. It is for this reason that we may consider that the main introduction of molecular simulation to Europe was made by Loup Verlet at the Orsay Science Faculty, built during the period 1960–1970 in the Vallée de Chevreuse near Paris.

So who was Loup Verlet? Why did his algorithm and his proposal for a list of nearest neighbours became so popular in the framework of our story? In order to answer these questions, we must take a step back, and return to the period indicated by the French physicist in the above quote, namely, the beginning of the 60s.

At the beginning of that decade, molecular simulation was a growing research field, mainly explored in the USA, since the technology required for molecular simulation, computers, was actually being developed in the USA. However, after about fifteen years of growth in computer power and refinement of the programming discipline (since 1945), everything was ready to export it to Europe. As mentioned above, an important point of entry (although not the only one) was Orsay, a small town with a few thousand inhabitants, 22 kilometres south-west of Paris. Here, in 1960, the Campus d'Orsay was being established, as a new branch of the "Faculté des Sciences" of Paris University, Sorbonne. As Dominique Levesque and Jean-Pierre Hansen report, in those months countless office buildings were erected, hosting laboratories for the whole range of sciences taught at the Sorbonne.[2]

One of these buildings hosted the Laboratoire de Physique Théorique et Hautes Énergies (LPTHE, Laboratory of Theoretical Physics and High Energy Physics). It was founded and initially directed by Maurice Lévy, a brilliant physicist, who was born in 1922 in Tlemcen (Algeria), and would be destined to a prestigious and multifaceted career. Indeed, within a decade, on December 19, 1973, Lévy was elected President of the Centre national d'études spatiales (CNES), a role he would keep until June 30, 1976. After taking part in the full pre-trial stage, a few years later, on May 21, 1985, Maurice Lévy became the first president of the *Cité des sciences et de l'industrie* for the diffusion of scientific and technical culture, the largest new-generation (i.e., interactive) science museum in Europe, situated within the Parc de la Villette in Paris, according to the specific wish of the French President Valéry Giscard d'Estaing. However, the museum was actually officially inaugurated by his successor, François Mitterrand, on March 13, 1986.

About 25 years earlier, in 1960, Maurice Lévy took charge of LPTHE in Orsay. It was a prestigious, but small centre, with at most thirty researchers, including PhD students, and occupied half a floor of building 211 of the Campus d'Orsay. The other parts of the building hosted the solid-state physics and plasma physics laboratories. Maurice Lévy had both vision and ambition. He wanted his laboratory

---

[2]D. Levesque, J.P. Hansen, *The origin of computational Statistical Mechanics in France*, European Physics Journal H **44**, 37, 2019.

to be multidisciplinary, and therefore recruited theoretical physicists already involved in various fields of physics, from nuclear to particle physics, and from statistical to mathematical physics.

The group studying statistical physics revolved around Loup Verlet, a first-class *normalien*, who had entered the *École normale supérieure* in Paris at the age of 17, a prestigious reference as far as the French scientific community was concerned. Indeed, Verlet was an excellent scientist. He might have built up a high-class research group and aimed at the Nobel Prize had he only gone all the way in the direction started in the 60s at Orsay, since he moved simulation up to a realistic level, thereby considerably helping all those who studied realistic systems, such as Martin Karplus, who won the Nobel Prize in 2013.

Loup Verlet (1931–2019) (photo Jean-Jacques Banide)

However, Verlet was not the kind of scientist to focus upon one question for his whole life. On the contrary, he was eclectic, involved in major sociological debates, and highly motivated to contribute to them. Thus he studied, in succession, nuclear physics and statistical physics, but also sociology, psychology, and philosophy. Each time, Verlet took up a subject in an all-encompassing way, while following an ideal path, which brought him, in the 70s, to abandon his strictly scientific activity.

However, we shall just stick to his activity as a physicist here. At the beginning of the 60s, Verlet was already well known for developing the HNC (Hyper Netted

Chain), a brilliant integral equation, devised for calculating the approximated radial distribution functions of fluid systems and their thermodynamic properties. This equation was paired with the one developed in the same period by Jerome Percus and George Yevick (known as the PY equation), which enabled significant theoretical advances in the development of the statistical physics of classical dense fluids.

In 1963, two papers, one by Everett Thiele and another by Michael Wertheim,[3] showed that the PY equation admitted an analytical solution for a fluid made up of hard spheres. Unfortunately, there was no such analytical solution for the HNC equation. The two approximate equations, PY and HNC, had similar shortcomings. For example, no analytical solution of these two equations could be obtained for realistic models of real fluids, such as those modelled by the interatomic Lennard-Jones potential.

At this point, Verlet had the idea of resorting to numerical solutions of these integral equations in order to compare the predictions of these approximate theories with experimental data. Since this comparison displayed clear discrepancies for dense fluids, Verlet derived systematic corrections to the PY and HNC equations, by including contributions of the so-called "bridge" diagram class, contributions overlooked by these approximate equations. To calculate these corrections, Verlet and his PhD student Dominique Levesque used the classical Metropolis Monte Carlo computational technique in 1962 to see whether the simplest "bridge" diagram contributed significantly to reducing the gap between the PY and HNC predictions, on the one hand, and the experimental results on the other. Moreover, for hard spheres, the predictions were also compared with the (exact) simulation results of Alder and Wainwright.

Solving the PY and NHC equations numerically and estimating the corrective terms proposed by Verlet had been made possible thanks to his intuition and foresight regarding the scientific role of computers, but also thanks to the joint efforts of Lévy, Verlet, and other physicists who had set up at the science faculty at Orsay in 1960–61 one of the largest and most advanced computer centres in Europe, equipped with several machines. One of them, the CAB 500, a French "drum memory" computer, designed by Alice Recoque and produced by the Société d'Électronique et d'Automatisme (SEA) was reasonably powerful. Indeed, it could carry out instructions and conclude arithmetical operations in about 40 ms. The SEA produced a hundred CAB 500 machines. The first was delivered in the month of February 1961.

However, the new centre at Orsay also disposed of an American computer, IBM 650, the first mass-produced computer in the world: two thousand of these were built. It was actually withdrawn from the market in 1962, but it was still cutting-edge at the time. Indeed, it carried out the same arithmetical operations with a speed an order of magnitude higher than CAB 500.

As we have seen, molecular dynamics and the Metropolis Monte Carlo method had obtained remarkable successes in the study of dense fluids, as is well documented in the literature. This fact did not escape the attention of Verlet's statistical

---

[3]E. Thiele, *Equation of State of Hard Spheres*, Journal of Chemical Physics **39**, 474, 1963; M. Wertheim, *Exact Solution of the Percus-Yevick Integral Equation for Hard Spheres*, Physical Review Letters **10**, 321, 1963.

physics group. In particular, the French group greatly appreciated the papers about hard spheres by Alder and Wainwright, since those papers helped to understand the way the PY equation succeeded in explaining the behaviour of that kind of system. On the other hand, problems arose when interpretating the comparison between theoretical results and experimental data from realistic systems such as rare gases, due to uncertainties regarding the validity of the intermolecular interaction model that was used in theoretical calculations, i.e., the Lennard-Jones potential.

Research studies by the LPTHE statistical group were carried out at a good pace until 1965, when Loup Verlet accepted the invitation by Joel Lebowitz and went to work for one year in the United States, at the Belfer Graduate School of New York. He was particularly intrigued by the paper *Correlation in the motion of atoms in liquid argon*, which had just been published by Anees Rahman and presented the simulation results, using molecular dynamics, of systems of particles interacting through the Lennard-Jones potential.

Verlet decided to extend this pioneering and seminal work on realistic fluids by the Indian physicist, thus granting it a more general dimension. He realized that molecular dynamics had not yet expressed its full potential, since not all interesting properties, such as, for example, the specific heat, had yet been calculated. However, in order to be able to do such computations, progress had to be made in the theoretical understanding of simulation.

Together with Lebowitz and Percus, Verlet was able to demonstrate how properties such as the specific heat could be obtained under various macroscopic conditions. In particular, at that time, molecular dynamics had been performed only under conditions where the energy was held constant, a macroscopic constraint on the system which specifies a microcanonical ensemble in statistical mechanics. In particular, they showed how in the microcanonical ensemble fluctuations in the kinetic energy could be used to calculate the specific heat of the system.[4]

It was by now clear that the conceptual turning points in molecular dynamics were accompanied—and often preceded—by a technological breakthrough. One of these breakthroughs took place just when Verlet arrived in the USA. The Courant Institute at New York University bought a new and powerful computer from the Corporation Data Company (CDC), and Verlet gained access to it. Loup Verlet fully seized this opportunity, developing and proving the validity of two closely linked ideas which were extremely important for molecular dynamics: the first concerned a stable and reliable algorithm which could be used to integrate Newton's equations of motion, while the second concerned the number of intermolecular interactions one had to compute in order to correctly calculate the evolution of the system.[5]

For additive pairwise interactions between n particles, this number is proportional to $n^2$, which is treatable, but also constitutes the most expensive part of the calculation. The innovation proposed by Verlet consisted in taking into account the relatively

---

[4]J.L. Lebowitz, J.K. Percus, L. Verlet, *Ensemble Dependence of Fluctuations with Application to Machine Computations*, Physical Review **153**, 250, 1967.

[5]L. Verlet, *Computer "Experiments" on Classical Fluids. I. Thermodynamical Properties of Lennard-Jones Molecules*, Physical Review **159**, 98, 1967.

short range of the interaction. This made it possible to cut the computing time for this calculation by a significant factor, in fact between 4 and 20. In other words, with Verlet's innovation, it took only a few minutes to complete calculations which had previously taken up to an hour of machine time. This was a great advantage and a significant breakthrough. Indeed, Loup Verlet's method is still used nowadays.

Naturally, Verlet's original molecular dynamics program was written in Fortran II CDC, i.e., in a language appropriate for the computer available at New York University. When Verlet came back to Orsay in 1966, he had to transcribe it into a language appropriate for the new computer at the science faculty there, namely UNIVAC 1107, and later on UNIVAC 1108. The latter was one of the most powerful computers available on the market between 1967 and 1970. As Dominique Levesque and Jean-Pierre Hansen report, the execution times for arithmetical operations on UNIVAC 1108 were on the order of a few microseconds, and moreover, the operating system was one of the first that could multitask, able to execute several programmes at the same time. For that reason many engineers at IBM France came to Orsay to take a closer look and observe the marvels of the UNIVAC machine and its innovative operative system.

We might therefore say that molecular dynamics was born in Europe at this date, 1966. And not only because this was when the statistical physics group in Orsay started to develop it in a systematic way, thanks to Verlet's ideas, but also, and above all, because Loup Verlet was one of the first European scientists to make an important creative contribution, giving new impetus to the development of molecular dynamics.

However, the activities of the statistical physics group at Orsay still retained their pioneering aspect. As such, they needed the full scientific creativity and determination of those who took responsibility for it. Indeed, UNIVAC 1107, and later on 1108, were not made available only to Verlet and his team. The computers belonged to the data centre used by the whole science faculty. Therefore, Verlet and his group could only get a certain share of machine time.

In any case, Verlet and his group soon gained visibility, in France and abroad. As Levesque and Hansen wrote: «It was hence on the basis of the equivalence, anticipated by Boltzmann, between molecular dynamics and the micro-canonical ensemble, combined with a heuristic implementation of practical details, that the use of molecular dynamics was developed at LPTHE in Orsay. It was, of course, recognized both at Orsay and elsewhere that a molecular dynamics program is readily transformed into a Monte Carlo program at the cost of a few minor modifications and the inclusion of a subroutine providing a truly reliable random number generator». This advanced and early awareness explains why, in the field of simulation, the statistical physics group at LPTHE started getting recognition, not only in Orsay and France, but all over the world.

To begin with, the group could count on their own group of brilliant researchers. Among these, beside Dominique Levesque and Jean-Pierre Hansen, we should mention Daniel Schiff, Jean-Jacques Weis, Jacques Vieillard-Baron, and several others. Moreover, the group benefited from an international network of collaborators, bringing together the best specialists in statistical mechanics, whether computational or non-computational: Ian McDonald, Konrad Singer, John Valleau, Georges Stell,

Mark Nelkin, Mal Kalos, Mike Klein, and Joel Lebowitz. And above an beyond these and everyone else, there were Berni Alder and Anees Rahman. This meant that Verlet had entered the restricted empyrean of the greatest scientists in the field of computational statistical mechanics. This is also why Verlet was by now considered one of the most brilliant and promising scientists in France. Indeed, in 1971, he was awarded the Paul Langevin prize of the Société Française de Physique.

However, between the end of the 60s and the beginning of the 70s, it was the whole group involved in statistical physics at the Campus d'Orsay that gained a reputation for, and a leading role in, computer simulation, for both molecular dynamics and Monte Carlo.

Jean-Pierre Hansen circa 1970

Dominique Levesque

The group set itself new and ambitious goals. Indeed, it was clear at this point, as Levesque and Hansen report, that simulation with both molecular dynamics and Monte Carlo, applied to simple, classical, or quantum systems, could not only establish the limits of validity of approximate theoretical approaches, but also predict new and unexpected physical phenomena, such as the solidification of hard sphere systems or the algebraic and non-exponential decay of correlation functions. However, it was much less clear whether these methods of simulation could be used effectively to study genuinely complex systems. This was the line of research which Loup Verlet wished to pursue, i.e., simulation of complexity, trusting in the continued exponential development of computer processing power.

In fact, at the end of the 60s, many highly diversified simulation projects were under way at the Laboratory of Theoretical High-Energy Physics in Orsay, using both molecular dynamics and Monte Carlo. Here is a short list proposed by Dominique Levesque and Jean-Pierre Hansen: Daniel Schiff was working on computer simulations to study simple fluid metals, based upon ion-ion effective potentials; Jacques Vieillard-Baron was the first to simulate systems of hard ellipses as a two-dimensional model of nematic order[6] and hence liquid crystals (this pioneering work inspired Daan Frenkel and his team in their ground-breaking study of the phase-diagrams of lyotropic liquid crystals[7]); Jean-Jacques Weis joined up with Berni Alder in Berkeley to work, also in collaboration with the experimental physicist Herbert Strauss, on the simulation of depolarized light scattering; Dominique Levesque and his team first applied molecular dynamics to simple models of rigid diatomic molecules; using both molecular dynamics and Monte Carlo simulations, Jean-Pierre Hansen, Ian McDonald, and Roy Pollock studied the static and dynamical properties of one-component plasmas (OCP), a model for highly-compressed plasma, generated by inertial confinement in nuclear fusion experiments.

Moreover, Verlet's group established close collaborations with other theoretical physicists at LPTHE, and in particular Bernard Jancovici, who had studied with Loup Verlet at the École Normale Supérieure. However, the group at Orsay, though well-known, ambitious, and well-connected with a good number of distinguished researchers, remained rather isolated. No other scientists in France were involved in computer simulations. Molecular dynamics was only being developed at Orsay. No matter. These studies were ground-breaking, since they were opening the way to the simulation of more and more complex realistic classical and quantum systems.

Towards the end of the 60s, to favour the development of the local science faculty, the leading French public research institute, the CNRS, established in Orsay the Centre Inter-Régional de Calcul Électronique (CIRCE), endowed with an IBM computer. This choice was at odds with the computational philosophy of both the faculty and Verlet's group, which had by now adopted UNIVAC. This led to a certain mistrust. IBM was not pleased about the fact that both the Laboratoire de Physique Théorique et Hautes Énergies and Verlet's group kept using UNIVAC. At the same time, physicists at the LPTHE, including the statistical physicists in Verlet's group, did not like IBM computers, and so did not frequent CIRCE and its structures, except on the odd occasion.

This detached attitude was not a good sign. Indeed, it was the prelude to the group's decline, which took place around 1974, when Loup Verlet and Daniel Schiff abandoned research in physics and started to become interested in human sciences, from philosophy to sociology; Jean-Pierre Hansen himself, though he kept close ties with Orsay, went to Paris, to the Université Pierre et Marie Curie. Thanks to Dominique Levesque, the work of the statistical physics group at Orsay went on, although they sorely missed Loup Verlet.

---

[6]Nematic indicates the class of liquid crystals for which the orientation of the molecules is ordered, but not their positions. A nematic liquid crystal flows like a liquid, but has similar optical properties to a crystal.

[7]Lyotropic liquid crystals are substances which, when dissolved in a suitable solvent, assume the typical ordering of liquid crystals.

Jean-Pierre Hansen, Dominique Levesque, Loup Verlet and Jean-Jacques Weis at the meeting in honor of Pierre Turq held in July 2014 at the Ecole Nationale Superieure de Chimie de Paris (courtesy Jean-Pierre Hansen)

Molecular simulation, however, had already spread elsewhere in Europe. At the Royal Holloway College near London, Ian McDonald and Konrad Singer were developing a very efficient simulation method for the study of classical fluids. In Austria, at the Technische Universität in Vienna, Kurt Binder was applying the Monte Carlo method to discrete spin systems, aiming to analyse neutron scattering data. At the Technische Universität in Munich, Bayern, Dietrich Stauffer started using the Monte Carlo method to study lattice percolation.

In the meantime, back at Orsay, on the hill overlooking the valley that hosted the statistical physics building of LPTHE, another research centre was set up. This was the Centre Européen de Calcul Atomique et Moléculaire (CECAM), destined, notwithstanding a certain scepticism and the ostentatious detachment of Loup Verlet, to become the major European centre of molecular simulation.

## 4.2 Britain: Konrad Singer and Ian McDonald

«We are grateful to Professor Verlet for providing us with a preprint of his paper». In December 1967, another small group in Europe, actually comprising just two people, namely professor Konrad Singer and his young assistant, Ian McDonald, two researchers at the Chemistry Department of the Royal Holloway College, University of London, published a paper in *The Journal of Chemical Physics*. The paper had the following title: *Machine calculation of thermodynamic properties of a simple fluid at supercritical temperatures*, and its abstract began like this: «A Monte Carlo method for the evaluation of equilibrium thermodynamic properties of a system of interacting particles is described».[8]

This was the last of a series of three papers the two scientists wrote about fluid simulation, in less than one year. The first paper had been published in January in the *Discussions of the Faraday Society* and concerned calculation of the thermodynamic properties of liquid argon using the Monte Carlo method.[9]

Konrad Singer and Ian McDonald were therefore the first scientists in the United Kingdom to attempt computer simulation, first using the Monte Carlo method, and later molecular dynamics.[10] However, those words of thanks to Loup Verlet in the paper published in December attest to the way, in the second part of the 60s, simulation was spreading to various parts of Europe, and was setting up a transdisciplinary community, at the same time interacting and distinct: Verlet, as we said earlier on, was a physicist, and his group was concerned with statistical physics; Konrad Singer, on the other hand, was a chemist who had up to then studied the nitration reaction in solution.[11]

Ian McDonald was a young Scottish physicist, born in England. He had just obtained a grant for a post-doctoral research position launched by Konrad Singer, who had himself been awarded a grant from the Science Research Council to work at the Royal Holloway College. With this award, he could pay a three-year position for a post-doc student capable of developing simulations.

When he started working at the Royal Holloway College, Ian McDonald knew nothing about simulation and the Monte Carlo method. «As for my state of mind at the outset of work with Konrad I can remember nothing. Nor did I know anything about work being carried out elsewhere, including France and the USA, nor of the uses in purely theoretical work to which simulation could be put. I was at the point of finishing my PhD thesis and simply needed a job to go to when I saw an advertisement

---

[8] I.R. McDonald, K. Singer, *Machine Calculation of Thermodynamic Properties of a Simple Fluid at Supercritical Temperatures*, Journal of Chemical Physics **47**, 4766, 1967.

[9] I.R. McDonald, K. Singer, *Calculation of Thermodynamic Properties of Liquid Argon from Lennard-Jones Parameters by a Monte Carlo Method*, Discussions of Faraday Society **43**, January 1967.

[10] In truth, a paper was published by E.B. Smith and K.R. Lea in *Nature* in May 1960 with the title *A Monte Carlo Equation of State for Mixtures of Hard-Sphere Molecules*, and this was followed 2 years later by another paper by the same authors, on the same topic. However, these two papers met with very limited success.

[11] Nitration reactions are chemical reactions involving the group $-NO_2$, derived from nitric acid $(HNO_3)$.

for a position in the chemistry department at Royal Holloway for work in what seemed to me to be an interesting field. I applied and was offered the job - end of story».[12]

Notwithstanding this confession of incompetence, Ian McDonald would become one of the leading actors in the development of computer simulation in Great Britain, publishing 37 papers in the years between 1967 and 1978 (seven of them with Singer) in which the Monte Carlo method and molecular dynamics played a central role. Therefore, Ian McDonald and his professor Konrad Singer should take the credit for opening the way to simulation in Great Britain.

The story of Konrad Singer belongs wholly to the tragic and somewhat epic period of twentieth century Europe. Singer was born in Bohemia in 1917, during the First World War, when Bohemia was still a part of the Austro-Hungarian Empire. He did not even realize that the Empire had broken up when the war ended in the month of November 1918. In any case, Singer was sent to school in Vienna, where he could nurture his two passions: music and science.

Apparently, he played the piano very well, and in due course entered the faculty of chemistry at the University of Vienna. He was attending his third year when, in 1938, Adolf Hitler decided the *Anschluss*, the annexation of Austria. Konrad was at a loss, observing the crowd cheering the *Führer* as he arrived in Vienna in triumph. He was working on his dissertation with professor Herman Mark, one of the pioneers of polymer chemistry. Singer's dissertation concerned the kinetics of the polymerization of styrene in solution.

Konrad Singer (1917–2013) (courtesy Peter Singer)

In the meantime, history imposed its conditions. Quite suddenly, Konrad's main problem no longer consisted in completing a good dissertation, but rather in leaving Vienna. Indeed, the race laws passed by the Nazis were also implemented in Austria. As a Jew, Singer could no longer attend the university.

---

[12]I. McDonald, private communication to Giovanni Ciccotti, July 2018.

However, for a short period of time, Jews were still allowed to leave territories under German control.[13] But in fact, other countries accepted Jews only if they already had a job, or, in the case of students, if someone would guarantee to pay for their studies. Konrad was lucky: he found a "sponsor" in England, namely, a generous *suffragette* pioneer of feminism and philanthropist called Marian Dunlop, who paid in part for his journey and for his PhD course. The rest of his expenses were paid by the International Student Service in Geneva, which was helping young Jews to leave Germany. Finally, the Scottish economist Alec Cairncross supported Konrad's entry to Glasgow University.

Thus in Scotland, in 1941, in the middle of the Second World War, Konrad Singer obtained his PhD, producing a dissertation on the optical properties of organic compounds. In the following years, Singer found a job in industry, and later on, in 1947, got a position at the Royal Holloway College of London, where he stayed until his retirement. Soon he met Jean Longstaff, who became his second wife in 1959. They had two children.

For a long time, until the end of the 50s, Konrad Singer had been interested in nitration reactions in solution. Later on, he began to study quantum chemistry. Then in the 60s, he moved on to the molecular simulation of fluids using classical techniques. At the time, this was a pioneering choice in England.

As already mentioned, from January 1967 McDonald and Singer published three papers in which they used the Monte Carlo method to calculate the thermodynamic properties of a Lennard-Jones model of liquid argon. This study was probably stimulated by experimental results on liquid argon which had just been obtained by two researchers at Oxford University. These results were published together with the third, more complete paper by Singer and McDonald in the same 1967 issue of the *Journal of Chemical Physics*. The third paper featured the use of the method known as "histogram weighting", or "re-weighting". It was a rather simple method, in which the scientist recorded the occurrences in which the potential energy of a system fell within a certain range. The data from the resulting histograms could then be weighted (in the microcanonical ensemble) or re-weighted (in the canonical ensemble) to obtain information about nearby states. This method would be redis-covered and improved about twenty years later, and is still widely used today in the analysis of data obtained through simulation.

---

[13]The epic of the Jews forced to flee German territories just before and during World War II has been told in an unpublished manuscript, found in an abandoned trunk in Konrad's house by his son Peter. The latter translated and published the book, written by Moriz Scheyer, an arts editor for one of Vienna's principal newspapers. The book tells the story of his adventures and contains detailed information about the fate of the Singer family: M. Scheyer, *Asylum: A Survivor's Flight from Nazi-Occupied Vienna Through Wartime France"*, Little, Brown and Company, 2016.

In 1969, Singer and McDonald published a paper entitled *Examination of the adequacy of the 12-6 potential for liquid argon by means of Monte Carlo calculations.*[14] It was essentially a reworking of a previous paper by Wood and Parker, published at the end of the 50s, in which it was shown, also on the basis of the new and highly accurate experimental data available at the time, that the 12-6 potential (i.e., the Lennard-Jones potential) was effective for argon. The method of histograms, previously introduced, was used in this case to optimize the parameters specifying the potential.

A few months later, Konrad Singer began to study the thermodynamic properties of binary mixtures of fluids, together with his wife Jean. At the same time, with his PhD student, Les Woodcock, he was still using the Monte Carlo method to simulate the behaviour of molten potassium chloride.[15] This work constituted another step forward for simulation methods. Indeed, it was the first time in which the Monte Carlo method was successfully used to study a molten salt, i.e., a system which involved explicit interactions between charges, whence its treatment required calculation of the effects of long-range interactions. The novelty of this work consisted in the expensive explicit calculation of "Ewald's sums". Indeed, the method, which took its name from the German physicist Paul Peter Ewald, provided a way to handle the long-range Coulomb interaction. The method had already been developed in the 60s in the USA by Brush, Sahlin, and Teller,[16] who used it to study the one-component plasma model, whereas Singer used it for the first time in the computer simulation of a realistic model of a two-component system. The paper published in 1971 immediately had a strong impact on physical chemistry studies of molten salts, and more generally, ionic liquids, and greatly affected the explosion of applications of molecular dynamics to realistic models of molecules, ionic systems, water, and solutions, which occurred over the following years.

The irony is that all this pioneering research was carried out by Singer's growing group at the Royal Holloway College, which did not even have a computer. Indeed, Singer, his wife, and McDonald had to fall back on the Atlas Computer Laboratory, established in Didcot in 1962. Didcot was a small town in the Thames valley, Oxfordshire, about 20 kilometres south of Oxford, and about 80 kilometres from London. The programme they used was written in the Atlas Autocode language, and of course used punched cards. Once the computer started to run in Didcot, it took it a few days to give results. Moreover, each time they had to change it, or even just correct some mistakes, they had to rewrite the whole programme.

---

[14] K. Singer, I.R. McDonald, *Examination of the Adequacy of the 12-6 Potential for Liquid Argon by Means of Monte Carlo Calculations*, Journal of Chemical Physics **50**, 2308, 1969.

[15] K. Singer, L.V. Woodcock, *Thermodynamic and Structural Properties of Liquid Ionic Salts obtained by Monte Carlo Computation. 1. Potassium Chloride*, Transactions of the Faraday Society **67**, 12, 1971.

[16] S. Brush, H. L. Sahlin, E. Teller, *Monte Carlo Study of a One-component plasma. I.*, Journal of Chemical Physics **45**, 2102, 1966.

In sum, calculation times were huge. The situation did not change much—apart from the travelling—when the Royal Holloway College acquired its own computer. This machine used the Fortran language (a step forward, but also a new investment in programming), which was not so fast, at least not at the time. And naturally, Singer and his team had to rewrite their programmes in the new language. At least, they no longer had to travel to and from London.

Notwithstanding the logistic difficulties, Konrad Singer's work continued to attract growing interest. This meant that he got access to more and more funding. Thus, at the beginning of the 70s, Singer expanded his research group, the first to work on computer simulation in Great Britain, which started using not only the Monte Carlo method, but also molecular dynamics.

Their field of studies was also expanding along new directions. In 1973, together with Ian McDonald and Eveline Gosling, Konrad Singer published in *Molecular Physics* a molecular dynamics simulation of the shear viscosity of simple fluids. This was one of the first attempts to introduce simulations of non-equilibrium molecular dynamics.[17] One of the first, demonstrating the fact that, when times are ripe, analogous developments happen more or less simultaneously and independently in different places. Indeed, Bill Hoover was working on the same kind of problems at Livermore.[18]

The collaboration with Ian McDonald went on for about twelve years, until 1979, even though their relationship took on a different nature. Indeed, the first grant expired after three years, and after that Ian McDonald obtained a lectureship in physics at Dublin University, in a department directed by the Nobel Prize winner Ernest Walton. Walton did not prevent Ian from continuing to cultivate his interests in simulation and working with Singer. On the contrary, it made it easier, since he paid his flights from Dublin to Heathrow, London's main airport. Thus McDonald continued to work with the simulation group at the Royal Holloway College. When Singer obtained new funds, he used them to further enlarge his research group and offered a substantial grant for a senior post-doc, at which point Ian moved back to London full time. As already mentioned, he stayed there until 1979, when he was offered a lectureship in Cambridge and a fellowship at Trinity College. The offer was made by phone, i.e., directly, without a call for candidates. It was an unexpected, tempting offer, «which, of course, I did not refuse!».[19]

---

[17]E.M. Gosling, I.R. McDonald, K. Singer, *On the calculation by molecular dynamics of the shear viscosity of a simple fluid*, Molecular Physics **26**, 1475, 1973.

[18]W.T. Ashurst, W.G. Hoover, *Argon Shear Viscosity* via *a Lennard-Jones Potential with Equilibrium and Nonequilibrium Molecular Dynamics*, Physical Review Letters **31**, 206, 1973; W.G. Hoover, W.T. Ashurst, *Nonequilibrium Molecular Dynamics*, Advances in Theoretical Chemistry **1**, 1, 1975.

[19]I. McDonald, private communication to Giovanni Ciccotti, July 2018.

Ian McDonald with Giovanni Ciccotti and Daan Frenkel in Amsterdam, 1986. They were
working at the volume "Simulation of Liquids and Solids", published by North Holland the
following year

Before all this, in the 70s, the interests of Singer and his group had extended
to cover the simulation of diatomic fluids. In 1979, for example, Konrad Singer
calculated the entropy of a molecule of chlorine ($Cl_2$) and bromine ($Br_2$), a property
which is much more difficult to obtain than a simple statistical average. Moreover
he developed an intelligent algorithm to integrate the equations of motion of a rigid
diatomic molecule well before the introduction of a method for handling the general
case.[20]

And this was still not enough. In 1972, McDonald showed that the techniques
of the Monte Carlo method could be extended to problems in which one maintained the pressure, number of particles, and temperature constant, rather than the
volume, number of particles, and temperature (canonical ensemble). His paper, published in *Molecular Physics*, described the first method to calculate—within the
isothermic-isobaric ensemble—the thermodynamic properties of binary mixtures of
liquids using the Lennard-Jones potential.[21]

Three years later, David Adams, another member of Singer's group at the Royal
Holloway College, made a further interesting step forward, by applying the Monte

---

[20]K. Singer, A. Taylor, J.V.L. Singer, *Thermodynamic and structural properties of liquids modelled
by '2-Lennard-Jones centres' pair potentials*, Molecular Physics **33**, 1757, 1977.

[21]I.R. McDonald, *NpT-ensemble Monte Carlo calculations for binary liquid mixtures*, Molecular
Physics **23**, 41, 1972.

Carlo simulation at constant chemical potential and hence a variable number of particles (grand canonical ensemble).[22]

In conclusion, the group that formed around Singer and McDonald was not only the first to introduce molecular simulation into Great Britain, but was also a pioneering team in the world as a whole due to the results it obtained. This process had consequences on the institutional level, too. In 1974, the first Collaborative Computational Project (CCP) was launched in Great Britain, to coordinate efforts in the various areas of the computational sciences at the national level. An important step in this context, where both Singer and McDonald were among the main actors, was the creation of the Collaborative Computational Project CCP5 in 1980, specifically devoted to the molecular simulation of liquids.

## 4.3  Austria and Germany: Kurt Binder

In the second half of the 60s, the third European centre focusing upon computer simulation in statistical mechanics came into existence in Germany, or rather in Vienna, Austria, where the 24 year-old Kurt Binder, who was studying at the Technische Universität for his PhD in physics, began to turn his attention to simulation. In his first paper, submitted in May and published in *Physics Letters A* in July 1968, Binder used the Monte Carlo method to calculate the spin correlation functions of a ferromagnet.[23] This paper was also signed by Helmut Rauch, an experimental physicist and Binder's "unofficial" thesis advisor. However, the computer programme had been entirely written by Kurt, who recalls his first approach to simulation: «In a seminar series on the physics of liquids, that I attended during the time my thesis work started I heard about the impact that the Metropolis method already had on understanding short range order (pair correlation functions, etc.) in liquids. At that time, there were no equally good competitive analytical methods for that purpose».[24]

As is well known, atoms in ferromagnetic materials tend to arrange themselves neatly, thanks to their spins, in the presence of an external magnetic field, or else, at low temperatures, through their so-called "exchange interactions". These interactions are a consequence of the Pauli exclusion principle, for the electrons carry magnetic moments that can align. However, although the basis of this interaction comes from quantum mechanics, Binder treated the magnetic moments quasi-classically, replacing spins by standard unit vectors. In this way, he obtained a model in the framework of classical statistical mechanics that exhibited a spontaneous phase transition.

---

[22]D.J. Adams, *Grand canonical ensemble Monte Carlo for a Lennard-Jones fluid*, Molecular Physics **29**, 307, 1975.

[23]K. Binder, H. Rauch, *Calculation of spin-correlation functions in a ferromagnet with a Monte Carlo Method,* Physics Letters **27**A, 247, 1968.

[24]This and the following quotations from K. Binder, private communication to G. Ciccotti, July 2018.

Binder, who was indeed a statistical physicist, wanted to get a deeper understanding of "how" the transition took place from normal disorder to this remarkably ordered state. A computer simulation immediately seemed to him to be the most suitable way: «Having the task to contribute to the understanding of spin correlation functions in magnetic materials, I had the more or less immediate idea that it should be interesting to try MC out for this problem. In the absence of better ideas, my advisor (Helmut Rauch) had no objection against this plan».

Kurt Binder was an ideal student, endowed with a remarkable insight and a pronounced independence. Therefore he decided to study the Monte Carlo method he had heard about, and apply it to problems he was concerned with: the transition from disorder to order in ferromagnetic materials.

It was no easy task. Nobody was specialized in simulation in Vienna. Kurt Binder was practically self-taught. He had to learn everything by himself, and also to do all the hard work alone. «I had to carry my deck of punched cards (I had written the code by myself, guided by the literature on MC for fluids, and punched the cards myself also) to the building of the computer centre, and even after the errors causing a failure of compilation of the program had been eliminated, I often found that the program had terminated, due to some computer breakdown».

Kurt Binder was more or less in the same situation as Berni Alder a decade earlier, and just like Alder, he learned to make do (and make sacrifices): «Since it seemed it was unpredictable how long the calculations would last in order to be successful, and since I always was an impatient person, I arranged with the operator of the computer that I could come by to his office and stay overnight with him. He slept there during the night on an emergency bed (I do not think that this was an official rule!) and I rather than him sat on the operator chair and watched the lights blinking ... When a breakdown occurred, I had to wake up the operator, so he could restart the machine again, and my four hour long runs could be started again ... I do not remember whether or not I had to bribe the operator (with a bottle of wine?) to allow me to regularly disturb his sleep ... But in this way I even managed to finish my thesis in a record time».

In any case, Kurt Binder's work was not just a thesis. It was actually a new line of research. In like manner, his second paper, sent in the month of October 1968, concerned the spin correlation functions of ferromagnetic materials.[25] Although alone and self-taught, Binder moved ahead rather quickly. In the course of 1969, he published two more papers. The last of these was signed by him alone.[26] By that time, Binder could be considered one of the pioneers of simulation in Europe.

[25]K. Binder, H. Rauch, *Numerische Berechnung von Spin-Korrelationsfunktionen und Magnetisierungskurven von Ferromagnetica*, Zeitschrift für Physik **219**, 201, 1969.

[26]K. Binder, *A Monte Carlo Method for the calculation of the magnetization of the classical Heisenberg model*, Physics Letters **30**A, 273, 1969.

Kurt Binder

In 1969, once he had obtained his PhD at the Technische Universität in Vienna, Kurt Binder moved to Munich, Germany to obtain the *Habilitation* (a qualification which would enable him to teach at universities) at the Technische Universität. He stayed in Munich for five years, until 1974, while continuing to use computer simulation to study ferromagnetic phenomena. During that time, he was the "unofficial" thesis advisor of two students who worked on pioneering applications of Monte Carlo simulation on ferromagnets, Volker Wildpaner in Vienna and Heiner Müller-Krumbhaar in Munich. While Wildpaner subsequently left research and went to work in industry, Müller-Krumbhaar later became one of the leaders of a theoretical physics group at the Jülich Research Centre in Germany.

When he obtained his qualifications in 1974, Kurt Binder refused an invitation from the Freie Universität in Berlin and accepted instead a position as professor of theoretical physics at the Universität des Saarlandes, a recent research university established only in 1948 in Saarbrücken, the capital city of this small Bundesland in West Germany, on the border with France.

In 1977, he married Marlies Ecker, and later on had two children. In the same year, he was chosen as leader of one of the theoretical physics groups at the solid-state research institute of the Forschungszentrum, Jülich, in North Rhein-Westphalia.

During this time, Kurt Binder gained a strong foothold as a pioneer of computer simulation working on the most relevant physical problems of the day, first in Austria and later in Germany. As with Verlet in France and Singer in England, he extended the scope of simulation, because he focused not only on ferromagnetism, but also, more generally, on phase transitions, glasses, and polymer physics.

Binder in Austria in 1968, Singer in England in 1967, and Verlet in France in 1966. Within less than two years, Europe had discovered computer simulation and had become a leader, together with the United States. The three research groups using the Monte Carlo method and molecular dynamics were almost entirely independent. These were symptoms that a general climate of attention to simulation was born and developing in Europe.

# Chapter 5
# CECAM and the Development of Molecular Simulation

## 5.1 Carl Moser and the Birth of CECAM

In the month of October 1969, thanks to the efforts of the American scientist Carl Moser, the Centre Européen de Calcul Atomique et Moléculaire (CECAM) was established in Paris, with a clear mission: «CECAM's purpose in the scientific world is to concentrate on problem areas on which only numerical solutions exist and for which larger and larger scale computing power becomes necessary for progress».[1] The research field on which the activity of Moser's centre would focus was not immediately defined. At the beginning, CECAM was working on almost everything. The idea of advancing in particular the studies of molecular simulation, and hence moving the best specialists in the world's scientific community to Orsay, would come a few years later. However, the establishment in Paris of the European Centre for Atomic and Molecular Calculation—which in 1994 would move to Lyon, and from 2009 onwards would be based in Switzerland—would mark a further significant step forward in the history of molecular simulation.

Carl Moser was a quantum chemist who was particularly interested in the forces acting between molecules. He was born on February 15, 1922 in Terre Haute, Indiana, and grew up in the nearby town of Clinton. He obtained his high school diploma in 1939, his first degree in chemistry at Indiana University in 1942, and his Master's degree in 1944. He worked in several private research laboratories (Allied Chemical, which later became Honeywell), where he was concerned with additives for tyres. Two years later, however, he decided to study for a PhD at the school of chemistry of Harvard University, where he could work with the well-known organic chemist, Louis Friser.

In 1948, Carl Moser presented his PhD dissertation on the properties of certain molecules used to fight malaria, a research line dear to Friser. In the following years, he published several papers on this topic, signing them with his colleague Marvin Paulsock.

---

[1] M. Karplus, *Carl Mathew Moser,* Carl Moser Symposium, CECAM, Lyon 2005.

© Springer Nature Switzerland AG 2020
G. Battimelli et al., *Computer Meets Theoretical Physics*, The Frontiers Collection,
https://doi.org/10.1007/978-3-030-39399-1_5

Thanks to these papers, he soon secured a position as a post-doc student at Cornell University and, immediately after that, a position as assistant professor in chemistry at Johns Hopkins University. As an organic chemist, he studied the synthesis of a long series of substitutes of benzoic acid, and the analysis of their UV (ultraviolet) spectra. However, Moser was aware that, in order to do well in this type of analysis, he needed a good knowledge of quantum mechanics, which he did not really have as yet.

Thus he asked his colleague Robert Burns Woodward, a well-known organic chemist, who would be awarded the Nobel Prize for Chemistry in Stockholm in 1965. Woodward suggested that he talk to—and possibly work with—an established theoretical chemist, Charles Coulson, who had pioneered the quantum theory of valence.

In 1951, Moser contacted Coulson and asked him whether they could work together. Coulson accepted and invited him to carry out research work with him. There was a catch, though: Moser was in the USA, whereas Coulson was English and worked in London until 1952, and later on in Oxford. However, Moser had no problem in crossing the Atlantic Ocean. Rather, that was a solution for him. Carl Moser was gay, and the political climate in the USA at the time was rather unpleasant for such people. Indeed, McCarthyism was raging, and according to a popular and thoroughly implausible argument, all homosexuals were communists, and therefore enemies of the country. In fact, in England, too, homosexuals were still being persecuted. Indeed, in 1952, Alan Turing, pioneer of computer science, was arrested for his homosexuality.

Perhaps Moser had not heard about that episode, but the fact is that he decided to cross the Atlantic Ocean, and move to Europe. At the same time, Moser also decided to abandon experimental organic chemistry and devote himself to theoretical chemistry, and quantum chemistry in particular. This area of chemistry was growing quickly, thus attracting the interest of a young and ambitious scientist like Moser. He started working on semi-empirical calculations on conjugated molecules, i.e., molecules in which simple bonds (exchange of one electron per atom) alternate with double bonds (exchange of two electrons per atom), and the delocalization of electrons is maximal. This topic was certainly very interesting. The American scientist started to develop his quantum skills. However, working with Charles Coulson was trickier than he had thought, because he seemed to have a rather difficult character. In conclusion, Carl Moser started to feel uncomfortable and impatient. He started looking for a chance to change once again.

Just like any typical American guy—Martin Karplus reports—Moser dreamt of spending Christmas in Paris. His dream was realized in 1952. In Paris, Carl Moser met Raymond Daudel, a professor of the Sorbonne who, in 1944, had founded the Centre de Chimie Théorique de France (CCTF), blessed by some of the most brilliant French chemists and physicists (Irène Joliot-Curie, Antoine Lacassagne, Louis de Broglie), whose aim was to apply quantum mechanics to the study of both chemistry and medicine. In 1954, CCTF was transformed into the Institut de Mécanique Ondulatoire Appliquée à la Chimie et à la Radioactivité (Institute of Wave Mechanics applied to Chemistry and Radioactivity).

Therefore, Daudel invited Moser without hesitation to take his leave from Coulson and join his staff at the new Institut de Mecanique Ondulatoire Appliquée, the first quantum chemistry group in France. The centre, he explained, belonged to the Centre National de la Recherche Scientifique (National Centre for Scientific Research or CNRS), the main French public research institute. Carl Moser thought about it, and accepted. In the fall of 1953, he crossed the Channel and settled in France, where he would remain until his death in 2004. He was literally an "American in Paris".

Carl Moser (1922–2004) with his beloved dogs in 1977 (CECAM archives)

For a while, Moser continued his studies on coupled systems with a semi-empirical approach. Later, he moved on to the ab initio analysis of systems of atoms and diatomic molecules: this was a field of study on the rise, and he could count upon one of the most powerful computers in the world, namely the one available at the computer centre in Orsay.

As mentioned above, Moser arrived in France when he was thirty, and so had no difficulties in moving into an apparently new and promising branch of science. On the other hand, as Martin Karplus wrote, Moser soon obtained a permanent position at CNRS—as research director—and therefore «was essentially free to do whatever he liked». Doubtless, Moser used all the freedom he was given in a creative and productive way. Indeed, in the month of October 1969, he set up CECAM, the Centre Européen de Calcul Atomique et Moléculaire. This centre, built upon international collaboration, promised from the very start to become a cutting-edge research institute in computational chemistry and physics.

CECAM was hosted in the same building as the Centre Inter-Régional de Calcul Électronique (Interregional Centre for Electronic Calculation or CIRCE), and could use CIRCE's IBM computer. Moser paid machine time to the CNRS, even though at a low "political" price, i.e., at a cost reserved for CNRS researchers. It is difficult to say whether this informality was an explicit choice, or just a lack of careful consideration about problems that were viewed as marginal. In any case, within the CIRCE building, no room was officially allocated to CECAM: there was only a gentlemen's agreement with Moser, according to which he would take up the top floor of the building. On the other hand, there was not even a sign at the entrance of the building with CECAM written on it to indicate where the centre was located. Moser and his research centre lived in a sort of underworld, without being seen. This was a deliberate choice, because Carl Moser feared for his own autonomy, and therefore carefully avoided raising the question of formalizing the centre.

Notwithstanding, the initiative was received rather coolly by Loup Verlet, who, at the time—let us repeat this once again—was France's leading authority in the area of molecular simulation. This coldness did not result only from the fact that Verlet's group, down in the valley, and CIRCE (therefore CECAM), up there on the hill, used two different computers and two different machine languages, but rather—and above all—because Moser wanted to build up a wholly independent institute, whereas Verlet resented the fact that CNRS funds would support other centres studying his own subject. On the other hand, Moser was reluctant to accept any external leadership. Moreover, he did not really appreciate the scientific skills of the French, and of Europeans in general. Coming from the United States, he certainly knew the difference in both leadership and infrastructures in the two scientific communities on the two sides of the Atlantic.

Verlet's coldness became open annoyance when Moser started to extend his field of interest from theoretical chemistry to Verlet's own area. They ended up in an all-out war, as CECAM began to get specifically involved in molecular simulation. However, over and above perceptions, actions counted. Indeed, the "American in Paris" started to connect with the same international network of renowned experts in molecular simulation established by Loup Verlet. This was the latest unacceptable impropriety for Verlet. Perhaps Verlet did not even realize—at least, not immediately—that Carl Moser had managed to get funds from some of these foreign scientists.

The first researchers who accepted to collaborate with Moser came from Italy, the Netherlands, and Belgium. Among the first Italians were Massimo Simonetta, a physical chemist from the State University of Milan, and Franco Bassani, a solid-state physicist and professor at the University of Rome "La Sapienza", with a wide international experience. He would soon be chosen as director of the Scuola Normale in Pisa. Thanks to Bassani, in 1972, another Italian researcher arrived in Orsay, Gianni Jacucci, a young experimental physicist from Rome who was interested in working specifically in the field of molecular dynamics, and would play a decisive role in guiding CECAM's activity towards that research area.

Summing up, through his quick and intelligent endeavours, Carl Moser acquired a well-equipped centre, obtained funds, and informally—without a prior written agreement—opened an account with the CNRS. Later on, with the funds brought informally by friendly foreign researchers, he supported their stay in Paris, together with their students. In short, he established a sort of reallocation of funds: he returned funds to their donors, while bringing them to his centre in Paris to carry out the research activity they were actually interested in. In this informal way, Moser obtained the participation in CECAM, not only of Italy, but also of the Netherlands, mainly through Herman Berendsen, and Belgium, through André Bellemans.

A nice picture of Herman Berendsen (1934–2019) and Carl Moser, taken in the early 70s

The participation of the United Kingdom took place in a slightly different way. Of course, many British scientists started coming to Orsay, but British institutes were more careful with their money, and stated that they could not support a centre with such an informal nature.

On the other hand, there were few German researchers who accepted the invitation of Carl Moser, since in Germany there were not so many specialists in molecular simulation. Kurt Binder was studying critical phenomena with the Monte Carlo method, but was not interested in CECAM, which only attracted a few German theoretical chemists who were in contact with Moser. Moreover the DFG, the equivalent of the Italian CNR and the French CNRS, was not authorized to grant international funding: this was a German application of the theory of the separation of duties. Indeed, the DFG was dedicated to research inside Germany, whereas German policy regarding international research was the sole responsibility of the appropriate ministries. The DFG could organize student exchanges and grants with foreign institutes. However, they could not handle common projects. At least, not all the time. Some might argue that West Germany was represented in international initiatives, such as the European Centre for Nuclear Research (CERN), but the DFG executives would answer that this was a foreign initiative which had been decided at government level.

In sum, notwithstanding all the difficulties we have mentioned, during its short life CECAM already had a strong and varied international network of collaborators. Therefore, there were now clearly two groups at Orsay: the only two groups in the whole of France that were involved in computational science, and they were glowering at each other. This was shown by the fact that, until Verlet studied molecular dynamics in the mid-70s, nobody in his group would take part in the more and more interesting initiatives of Moser's CECAM.

The fate of the two adversaries would soon diverge. As we shall see later on, CECAM would prevail in the end. On the other hand, Verlet's group slowed down after its founder, Loup Verlet, left in the mid-70s to pursue new cultural interests. After all, Verlet and his group had some powerful adversaries, such as Pierre-Gilles de Gennes (who would be awarded the Nobel Prize in 1991), who was a brilliant scientist and also an accomplished politician. The fact is that de Gennes started fighting effectively against Verlet's group which, for its part, was badly affected by the loss of its former leader. And so France missed an opportunity there.

Let us go back to CECAM. The arrival of Gianni Jacucci turned out to be a crucial factor, a genuine asset for the new centre, enabling it to achieve a really ambitious goal: to focus on and take a leading position in the field of molecular simulation. Indeed, following the arrival of the Italian scientist, the thematic workshops that had already been initiated in 1972 began to play «an important role in the development of molecular dynamics, and, in particular, in its application to biological macromolecular systems».[2]

The first of these meetings, the workshop on *MD and MC calculations of water*, which focused on the study of water, using both Monte Carlo and molecular dynamics, had actually been organized before Jacucci's arrival. The idea of this specific workshop had originated when Carl Moser met two Dutch colleagues, Wim Nieuwpoort and Herman Berendsen, in Groningen to discuss CECAM's activities, which up to then had concentrated upon quantum chemistry.

---

[2]H.J.C. Berendsen, *The Development of Molecular Dynamics at CECAM*, in J.P. Hansen, G. Ciccotti, H.J.C. Berendsen, *In Memoriam. Aneesur Rahman. 1927–1987*, CECAM, Orsay 1987, pp. 9–12.

In 1972, Rahman and Stillinger had published their paper on the molecular dynamics of liquid water.[3] Then the idea arose to organize at Orsay, in a year's time, a meeting devoted to the applications of Monte Carlo and molecular dynamics to the study of water. «It was immediately clear—Berendsen recalls—that in such a workshop Anees Rahman should be the key scientist, whose knowledge—and programs, we hoped—would be available for the other participants».

Rahman—whom Moser knew well, since he had been a guest at Orsay in 1971—was thus presented with the idea of the workshop. Rahman had doubts at first, but was soon convinced. The meeting could go ahead and was soon organized. Anees Rahman, who loved Paris and went there often, was the central speaker of the workshop. Above all, the meeting was just the first stage of a collaboration established by Rahman with CECAM, where he undertook to stimulate the study of molecular dynamics of systems constituted of water, simple fluids, molten salts, solids, and eventually also biological macromolecules.

At the workshop, Rahman talked first, and presented the program on the computer with which he had first simulated the molecular dynamics of liquid water. Rahman's talk was very difficult to understand, because, as Berendsen recalls, «his programs and output listings were quite unintelligible for others: they did not contain a single comment or heading, indeed only numbers appeared on the output listing. He was extremely pragmatic and would not spend one moment on what he considered irrelevant». Indeed, his presentation was full of formulae, with only a few qualitative explanations. Rahman was perfectly aware of this: he explained that it was just a matter of numbers, because all the rest was irrelevant. Hermann Berendsen, however, criticized him regarding both the content and the form, and told Rahman that his computer program was not the best that could be made. However, Rahman's simulation had one great asset: it worked and it contained no bugs.

In any case, the first CECAM workshop was not confined to Rahman's talk. Rather, it focused upon open problems in the molecular simulation of liquid water, some of which had a general character, like the appropriate handling of long-range molecular interactions, whereas other problems were more specific, such as the connection between dipole fluctuations and the dielectric constant or the polarizability.

However, over and above the analysis of individual problems, the workshop opened up a far wider outlook on things. Once again, Berendsen recalls: «The 1972 workshop showed that simulation of water was possible. Why not more complicated systems?». This idea was certainly not to advocate a head-on pursuit. Indeed, water already contains the main difficulties of long-distance polar interactions and polarizability, whereas its behaviour, largely determined by the hydrogen bond, is crucial in biological macromolecules. «If you cannot simulate water—Berendsen explained—biological macromolecules will be hopeless; if you can, the macromolecules may turn out just to be more complex, but not more complicated».

---

[3] A. Rahman, F. Stillinger, *Molecular Dynamics Study of Liquid Water*, Journal of Chemical Physics **55**, 3336, 1971.

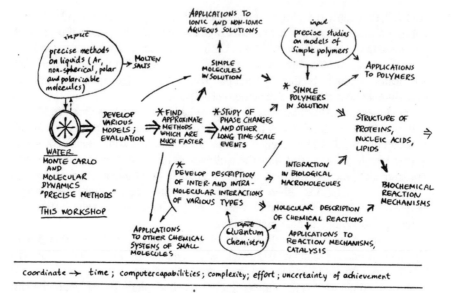

Herman Berendsen's prevision for the future developments of molecular simulation, that he presented at the 1972 CECAM workshop on "Molecular Dynamics and Monte Carlo calculations of water"

In sum, the first CECAM workshop opened up a window on an unprecedented and very important part of computer simulation. This area was explored by the simulation of the dynamics of biological molecules, including macromolecules such as proteins and nucleic acids. The project was a good one, provided that someone could give it substance and continuity. With the arrival of Gianni Jacucci, all the prerequisites were in place. Indeed, Jacucci seemed fully capable of taking up the suggestions of the workshop, thus contributing to the rapid development of this ambitious project.

## 5.2   Molecular Simulation Between Europe and America

On the other hand, the time was now ripe for this kind of work. This was confirmed by the fact that, in the same year, 1972, the *Journal of Physics* published a paper by two English researchers from the University of Manchester—Arthur W. Lees and Samuel F. Edwards—on *The computer study of transport processes under extreme conditions*, reporting on research in which molecular dynamics had been used to simulate the behaviour of a fluid under the action of an extremely strong shear force.[4] There were still problems, the two English researchers said, for the accurate calculation of the

---

[4]A.W. Lees, S.F. Edwards, *The computer study of transport processes under extreme conditions*, Journal of Physics C **5**, 1921, 1972.

viscosity. To this end, the computing power had to be increased. However, even a short computation was enough to establish the nature of the processes involved and obtain the velocity distribution curve. Lees and Edwards showed these charts in the paper. Their method, they stated, allowed them to abandon the periodic boundary conditions which had been used up to then in molecular dynamics calculations, and replace them with more general boundary conditions.

It was clear that the computer simulation of molecular dynamics in the early 70s was evolving quickly, and after its early stages, was proving to be a very powerful and flexible tool with which scientists could explore the dynamics of dense fluids and/or liquids, under any conditions.

While molecular simulation was taking its first crucial steps in Europe, extending the spectrum of features of natural systems that could be investigated, things were not at rest in the USA. Among others, Berni Alder was actively carrying out his studies. As mentioned above, in 1971 his friend and colleague Tom Wainwright left the study of molecular dynamics to take up the direction of Livermore's Theoretical Physics Division. Two years later, Wainwright obtained the Ernest O. Lawrence prize for his exceptional contributions to the research and development of the theory of liquids, and in particular transport properties and hydrodynamics, in computational physics. The following year, 1974, Wainwright became a fellow of the American Physical Society.

Let us therefore go back to Berni Alder. In 1972, in the third volume of a journal— *Computer Physics Communications*—founded three years earlier by Phil Burke at Belfast University «to facilitate the exchange of physics programs and of relevant information about the use of computers in the physics community», Alder published a conceptually strategic paper, *Numerical experiments in statistical mechanics*.[5] In this paper, Alder discussed the potential and the limitations of both the Monte Carlo and molecular dynamics methods for the study of classical and quantum systems. In his view, the potential and results achieved by molecular simulation in the study of classical systems, even in the case of non-equilibrium, had by then become clear. The great novelty, according to Alder, was that computer power had reached such a point that it allowed one to consider simulation, not only of classical Newtonian systems, but also of quantum systems, through the numerical solution of Schrödinger's equation. He said that several problems remained to be solved, but it was finally possible to get started.

Here, therefore, was the new frontier: the simulation of quantum systems. Complementary to the one indicated at Orsay: the simulation of the molecular dynamics of biological systems.

Alder's paper began with a statement about the effectiveness of molecular dynamics in addressing and solving classical non-equilibrium problems. «The computational problems in statistical mechanics», he wrote, «can be divided into four separate areas each requiring special numerical procedures. Most of the work so far has been confined to classical problems. For the equilibrium properties the so-called

---

[5]B.J. Alder, *Numerical experiment in statistical mechanics*, Computer Physics Communication **3**, Suppl. 86, 1972.

Monte Carlo procedure is employed. For non-equilibrium properties the molecular dynamics method, which simply solves the simultaneous Newtonian equations of motion, is used, although this procedure yields equilibrium properties as well and with comparable efficiency to the Monte Carlo procedure».[6]

Then Alder went on to discuss quantum problems: «The quantum mechanical versions of these two classical methods require much more effort. For equilibrium properties, and in the case when only a few energy levels, or particularly the ground state energy is required, the variational method can be employed. It is frequently used in chemical many-body calculations, but the accuracy requirements in that case are so high that it cannot be at present considered a reliable method when more than 3 electrons are involved». Nevertheless, Alder went on, the Monte Carlo method could also sometimes be used, with a few approximations, for quantum systems with hundreds of particles, such as liquid helium, or a gas of electrons. On the other hand, it was still too early to simulate the dynamics of quantum systems involving many particles. A useful path might consist in transforming Schrödinger's equation into a hydrodynamic equation through a suitable change of variables. By then, indeed, it was known how to solve hydrodynamic equations for a single particle moving in an external potential. This ability could now be extended to many-body systems.

A year later, in 1973, Alder published *Computer Dynamics*, a state-of-the-art review of molecular simulation, where, on the one hand, he emphasized the advantages of computer simulation in comparison with both approximate theoretical and expensive experimental approaches, and, on the other hand, he underlined its current limitations and indicated a possible future path: «The aim is to introduce ever more realistic potentials in these calculations to reproduce experiments on ever more complex systems».[7]

Oddly enough, Alder did not mention the papers published by Anees Rahman (1964) and Loup Verlet (1967), whereas he referred to the paper published by Rahman and Stillinger in 1971 on the simulation of water—even though, he wrote, these studies were still «in their infancy»—and also to the more recent papers by Verlet and his team. It is difficult to explain this omission. Perhaps he was simply not used to handling the problems addressed by Rahman and Verlet, namely, computer simulation of continuous realistic potentials. Or else, there may be a deeper reason. As he would say later on, at the beginning of the 70s he had been focusing on more fundamental problems: he wanted to provide a solid foundation for the methodology of molecular simulation. That was indeed why he had turned to quantum mechanics. He was convinced that, unless one got to the quantum level, any computer simulation of real systems, such as water molecules, would necessarily be incomplete. Therefore, he preferred to put the emphasis on the quantum Monte Carlo method.

---

[6]In fact, here Alder was thinking that, thanks to a theorem of non-equilibrium statistical mechanics, the non-equilibrium properties might be obtained by computing equilibrium time correlation functions. Using this method, in his papers on transport, Alder could calculate non-equilibrium properties from dynamical equilibrium properties.

[7]B.J. Alder, *Computer dynamics*, Annual Review of Physical Chemistry **24**, 325, 1973.

Meanwhile, as mentioned earlier on, in the journal *Molecular Physics*, in 1973, Evelyne Gosling, Ian McDonald, and Konrad Singer proposed an authentic "computer experiment", and obtained interesting results from the molecular dynamics study of the shear viscosity of a simple fluid.[8] This was an example of a growing application of computer simulation to the area of transport properties. The logic of simulation was expanding.

At CECAM, they were beginning to realize the importance of the change which was taking place. Indeed, in 1973, the first CECAM Discussion Meeting was held at the Warffum castle in the Netherlands to establish future research trends. Among other things, it was decided that two workshops would be held in 1974, one on ionic fluids, organized by Konrad Singer, and the other on long time-scale events, in order to face possible problems coming from the molecular simulation of biological molecules.

In particular, molecular dynamics at CECAM was developing in two directions that were not always followed by the same people, but were not completely separate: on the one hand, the "exact" handling of simple systems (simple fluids, ionic liquids, the behaviour of non-equilibrium systems), and, on the other, the approximate handling of models of the real world, in an attempt to understand and predict the behaviour and properties of complex real systems. This was the case for biological macromolecules, such as proteins and DNA.

Moser's institute soon became a cutting-edge research centre in all these fields, in particular in the study of molecular liquids and molecules of biological interest. So much so that CECAM came up with the molecular dynamics of proteins, in particular proteins in water solutions. In the course of one of the two workshops, held in 1974, organized by Herman Berendsen, a lot of ideas were actually developed for the study of the molecular dynamics of biological molecules, and that set the stage for a ground-breaking meeting that took place two years later, which we shall discuss in detail further on.

The other 1974 workshop, organized by Konrad Singer, devoted to the *Study of Ionic Liquids by Computer Simulation*, was attended by most of the leading figures of molecular dynamics of the time, such as Anees Rahman, Harold Friedman, Mario Tosi, Ian McDonald, and others, including some of the long-term CECAM visitors. In particular, the workshop saw the arrival at CECAM of Mike Klein, another relatively recent newcomer to molecular simulation, destined soon to become a leading figure in the field. Born in London in 1940, Klein had a background as a theoretical chemist. He had obtained his PhD in Bristol and then crossed the Atlantic to Rutgers, before finally settling in Canada in 1968, where he was entrusted with the creation of a computational chemistry group in Ottawa, under the aegis of the National Research Council of Canada (NRCC).

In Ottawa, Klein learned about molecular simulation, and was particularly impressed by the Monte Carlo simulation of water by Barker and Watts. Early in 1970 he visited Barker at IBM in California, and a collaboration started which had

---

[8]E.M. Gosling, I.R. McDonald, K. Singer, *On the calculation by molecular dynamics of the shear viscosity of a simple fluid*, Molecular Physics **26**, 1475, 1973.

as an outcome a joint paper on the Monte Carlo computation of the melting line of argon, published in 1973.[9] Meanwhile, he became aware of the pioneering molecular dynamics studies by Rahman on argon and water, and started working on the comparison of Monte Carlo results with the self-consistent phonon theory (his own expertise as a solid state theorist) together with Bill Hoover. The real moment of conversion to simulation, with a shift of focus from Monte Carlo to molecular dynamics, came with his encounter in the US with Jean-Pierre Hansen in 1970. This fateful and entirely accidental contact generated several visits to Paris and an acquaintance with the members of Verlet's group in Orsay. Together, Hansen and Klein published a first paper on the dynamical structure factor in 1974, to be followed soon after by others on the same general subject, studying different realistic model systems.[10] At the same time, attending the 1974 workshop on ionic liquids, Klein came in contact with the CECAM group, and in particular, apart from Anees Rahman, with Gianni Jacucci and Ian McDonald. A strong collaboration was established, leading to a number of joint papers starting in 1975.[11] In the course of a sabbatical year spent in Paris in 1975–76, he also had the chance to meet Berni Alder, also visiting at Orsay, and another joint work emerged from the ensuing collaboration.[12]

From that moment on, Klein developed an amazing ability to master all sorts of new simulation techniques and computational methods, applying them to the widest range of model systems. «He has proven himself a true master at using simulation to make models come to life, produce results that can be compared directly to experiment, and help us to understand what is going on in the "real world" (one of his favourite expressions)… If not making the first step in the development of a new method, then he often made the crucial second step of applying it to the "right" system, showing what the method could contribute to solving problems of interest to theorists and experimentalists alike».[13] Klein was instrumental in raising a flourishing group of computational scientists in Ottawa, often creating the conditions for

---

[9]J.A. Barker, M.L. Klein, *Monte Carlo calculations for solid and liquid argon*, Physical Review B **7**, 4707, 1973.

[10]One of the first papers to appear in print was J.P. Hansen, M.L. Klein, *Computer "experiments" on solid rare gases: the dynamical structure factor S (Q, ω)*, Journal de Physique Lettres **35**, 29, 1974. It contained only a limited test of the validity of self-consistent phonon theory vs molecular dynamics. Instead, by the summer of 1973, using the Univac computer at Orsay, Hansen and Klein had produced a full study in which molecular dynamics was used to test the semi-analytic theory; at the time a real comparison with experiment was not possible, because the force field used was too primitive. The paper had been submitted to Physical Review Letters, but difficulties with the referees delayed publication until 1976, when an extended version was completed during Klein's sabbatical year, spent in Paris in 1975–76: J.P. Hansen, M.L. Klein, *Dynamical structure factor S (Q, ω) of rare-gas solids*, Physical Review B **13**, 878, 1976.

[11]G. Jacucci, M.L. Klein, I.R. McDonald, *A molecular dynamics study of the lattice vibrations of sodium chloride,* Journal de Physique Lettres **36**, 97, 1975.

[12]B.J. Alder, H.L. Strauss, J.J. Weis, J.P. Hansen, M.L. Klein, *A molecular dynamics study of the intensity and band shape of depolarized light scattered from rare-gas crystals*, Physica B+C **83**, 249, 1976.

[13]M. Sprik et al., *Tribute to Michael L. Klein: Scientist, Teacher, and Mentor*, Journal of Physical Chemistry B **110**, 3451, 2006.

collaborations among people working on contiguous fields. In the next chapter, we will discuss the development of molecular simulation over the following decade, giving rise to a real explosion of methods and ideas in the most varied directions; right through this period, almost all the creators of the innovations were in some way or other related to Klein, either collaborating directly with him or interacting with his research group, first in Ottawa and later, from 1987 on, in Philadelphia.

Mike Klein with his wife Brenda, circa 1973

Let us go back to CECAM and the 70s. As mentioned earlier, its main feature was its workshops focusing on various topics, not necessarily ordered in a homogeneous manner: quantum chemistry, plasma physics, atomic physics, and condensed matter physics. Each meeting was organized by one or more experts, in general friends of Moser. After the arrival of Jacucci, the activity of the workshops on molecular simulation soon exploded, making CECAM, up there on the hill, an even more active simulation centre than Verlet's, down there in the valley. Indeed, Moser's workshops were not only organized to meet and talk, with the idea of solving everything in a few days. For its participants, CECAM offered collaboration with young researchers on the spot, as well as machine time for a few weeks, so that these workshops were a genuine opportunity to carry out pioneering work. This new arrangement was truly successful, mainly with researchers from small countries which did not have a powerful computer centre like the one offered by Carl Moser.

On the other hand, CECAM lacked organization. Moser had a full-time secretary, as well as a part-time secretary. Though they both worked for him alone, that was all. At CECAM in those days, it could happen that nobody answered the phone. Formal organization was definitely not its strong point.

As mentioned above, in 1972 the Italian researcher Gianni Jacucci arrived at Orsay. Until then, he had been concerned with the spectroscopy of fluids, but he had heard about molecular dynamics. He was very interested in the new topic and knew of a cutting-edge group in France, namely Loup Verlet's group, with whom he would like to work. However, he finally chose CECAM, following a recommendation by Franco Bassani. Indeed, only Moser's centre could support his long stay.

As we said earlier on, Verlet did not really like CIRCE, let alone CECAM. He did not appreciate Carl Moser's scientific worth. He did not want to deal with those uncouth people up there on the hill. In any case, Gianni Jacucci was not interested in all that. He was only interested in molecular dynamics. He collaborated with Verlet's group, but at CECAM too, he was soon focusing on computer simulation. In fact, he soon literally got hold of the machine. From 6 a.m. on a Sunday until 6 a.m. on a Monday, CIRCE left its IBM completely in his hands. Not even the keeper was there, since this was their weekly day off. Clearly, Gianni Jacucci sacrificed all his Sunday activities, and spent the whole day and the whole night on the computer, trying to make it work.

Jacucci's arrival marked a new phase for CECAM, because he achieved something that Moser could not do: direct the centre towards specific, focused, and highly-productive activities, thanks also to the help—and the presence—of high-value scientists, such as Anees Rahman and Martin Karplus, who were based in the USA, but often came to Europe, even for long periods. In this respect, Carl Moser really played a key role, because he offered them a natural point of contact with Paris, namely, the workshops and the availability of a powerful computer.

However, Gianni Jacucci himself was able to take advantage of the presence of scientists such as Rahman and Karplus, and remarkably sped up CECAM's work in the field of molecular simulation, all the while maintaining the delicate balance with Loup Verlet.

The difficulty and fragility of this balance can be illustrated by an anecdote. Jacucci used a program made by a collaborator of Verlet, Dominique Levesque. Thus, every time he published a paper, Jacucci thanked Levesque. Once—only once—he forgot to do that. Loup Verlet heard about this and, although he had abandoned his research, he wrote a letter to Jacucci, and called him «a shark».[14] The matter was solved through the mediation of both Rahman and McDonald, who were there at Orsay for a workshop. Loup Verlet finally accepted Jacucci's apologies.

In the meantime, in March 1974, Giovanni Ciccotti arrived at CECAM. He had studied at the University of Rome and was a classmate and friend of Gianni Jacucci. When he arrived in Paris, invited by Jacucci, he was just over thirty years of age.

---

[14]Personal recollection of G. Ciccotti.

Shortly after that, Jacucci and Ciccotti published a paper together, in which they presented a significant extension of simulation techniques.[15] Since 1973, scientists had started to consider the possibility of using molecular dynamics simulations not only in conditions of equilibrium, but also under stationary external perturbations. Bill Hoover at Livermore Laboratories and Konrad Singer and Ian McDonald in London were the first scientists to address this problem. Jacucci and Ciccotti carried the torch, and found a mathematically rigorous way to calculate averages, even in conditions of dynamic non-equilibrium, for perturbations which are not only stationary, but also explicitly time-dependent. Moreover, by introducing a numerical expedient, indeed, only useful for short times, they also managed to reduce the statistical noise and see what happens with small perturbations. This method, called the "subtraction technique", was immediately noted and mentioned, even though not entirely understood. Indeed, it would be used for years in its most elementary form, and was then practically abandoned. Recently, physicists have realized that this method may have a more general application, and it has thus been proposed for carrying out hydrodynamic simulations.[16]

Jacucci stayed at Orsay and worked at CECAM for five years, during which he reinforced the team, mainly with young people, willing to work hard on everything he proposed. Moreover, he extended simulation to other centres, such as the French atomic energy authority (the Commissariat à l'énergie atomique or CEA). In sum, he provided strong roots for the rather delicate sapling that Moser had planted at CECAM.

The workshops, of course, continued. Even though, over time, not all of them kept up the brilliant standards of the first. With all the advantages and disadvantages of informality, Carl Moser certainly created a valid scientific structure, which Gianni Jacucci directed towards molecular simulation. CECAM kept growing and remained successful, at least until 1980, when the Italian CNR, pushed by union concerns, raised a question: why are we giving money to a structure that does not even exist? Thus the top Italian public research institute pulled out. Carl Moser was on pins and needles. Indeed, this could be the beginning of a chain reaction. Moreover, Gianni Jacucci had left the centre three years earlier. Indeed, in 1977 he went back to Italy in a huff, because Moser had never thought of finding a permanent position for him. Probably, the American scientist did not want a leader who might overshadow him in his own centre, or could in any case exert pressure. It is noteworthy that, looking around for a place to come back to Italy, Jacucci, who had started out as an experimental physicist, found it difficult to gain recognition in Rome as a simulation theoretician, either by theoreticians or by experimentalists, notwithstanding his undeniable role as an innovator at Orsay. He therefore abandoned the University of Rome and moved to

---

[15]G. Ciccotti, G. Jacucci, *Direct Computation of Dynamical Response by Molecular Dynamics: The Mobility of a Charged Lennard-Jones Particle*, Physical Review Letters **35**, 789, 1975.

[16]S. Orlandini, S. Meloni, G. Ciccotti, *Hydrodynamics from dynamical non-equilibrium molecular dynamics*, AIP Conference Proceedings **1332**, 77, 2011; G. Ciccotti, M. Ferrario, *Non-equilibrium by molecular dynamics: a dynamical approach*, Molecular Simulation **42**, 1385, 2016.

Trento. This episode tells us how difficult it was for the Italian academic community to appreciate novelty if it jeopardized well-established disciplinary boundaries.

In the five years Jacucci spent at CECAM, up until 1977, there were simulations of molecular dynamics for non-equilibrium systems and also successful applications to biology, which actually witnessed a genuine boom. The research concerned not only dynamics simulations, but also a number of other questions, such as the relation between the structure and function of biological macromolecules. Soon the need and the opportunity emerged to study the folding of proteins and DNA, enzymatic reactions, and the electrostatic fields which surround large biological molecules.

A turning point came in the month of February 1975, with a paper by Michael Levitt and Arieh Warshel, who in 2013 would be awarded the Nobel Prize for Chemistry, together with Martin Karplus, for their studies on the development of multi-scale models for complex chemical systems. This paper, entitled *Computer simulation of protein folding*, constituted the first attempt to simulate on a computer the folding of protein macromolecules.[17] The paper showed that, under certain conditions, the simulation could "regenerate" bovine pancreatic trypsin, thus moving from an open-chain structure to a folded structure, similar to that of the native molecule.

Martin Karplus (photo by N. Pitt 9/10/03)

---

[17]M. Levitt, A. Warshel, *Computer simulation of protein folding*, Nature **253**, 694, 1975.

For the moment, however, this was a rather rough, although stimulating, simulation. Indeed, in 1975, a new CECAM discussion meeting debated upon three alternatives: focusing on very accurate simulations of simple systems, focusing on rougher simulations of biological macromolecules, or merge the two approaches. This question originated a new workshop the following year, on *Models for Protein Dynamics*, a real milestone in the simulation of large molecules of biological interest.

## 5.3 Towards Biology: The Simulation of Proteins

It is often said that in the final part of the twentieth century biology undermined the primacy of physics among all sciences. It is indeed difficult to establish such hierarchies. No doubt in the last four decades, starting from the mid-70s, studies in biology—especially at a molecular level—speeded up remarkably. This boost was mainly due to investigation into the dynamics of proteins and, more broadly, the dynamics of large biological molecules. This development was also helped by CECAM, which, in 1976, organized a workshop on this topic at Orsay.

Forty years later, in 2016, Emanuele Paci, a researcher at the Astbury Centre for Structural Molecular Biology at the University of Leeds, organized a meeting in order to celebrate that workshop. The meeting, entitled *Models for Protein Dynamics, 1976–2016*, was held from 15 to 18 February 2016 in Lausanne, the new CECAM headquarters. In his introduction to the workshop, Paci wrote: «The role that modelling and computation has played in our understanding of how proteins and biomolecules work has been steadily increasing in the last 40 years». He went on: «Biomolecular simulation has been instrumental in revealing the role of dynamics in biological function, and in directing the development of experimental techniques more suited to understand biological function, such as single molecule ones». Finally, Paci stated: «CECAM has played a crucial role in the development of a steadily broadening field starting from a landmark workshop held in Orsay from May 24 to July 17, 1976».[18] Emanuele Paci here refers to the workshop organized, according to the CECAM tradition, over eight weeks between May and July 1976, by the Dutch biochemist Herman Berendsen, of Groningen University, and entitled *Models for Protein Dynamics*.

In the mid-70s, many scientists were discussing whether it was possible to study the molecular dynamics of large biochemical molecules, in particular proteins. Herman Berendsen recalls this period: «We organized a huge two-month workshop on

---

[18]E. Paci, *Foreword*, unclassified document, circulated among the participants of the 2016 workshop, with contributions by M. Karplus, W. van Gunsteren, J.A. McCammon, S. Wodak, and G. Ciccotti, CECAM 2016.

*Models for Protein Dynamics* with 22 participants. In practice, the accurate simulators and the crude biophysicists formed two groups with highly deviating interests
and preferences. But there was a lot of interaction and some could bridge the gap».[19]

Anees Rahman, by now unanimously recognized as one of the fathers of molecular
dynamics, was among those participants who could bridge the unprecedented gap
between simulators and biophysicists. Rahman, as Berendsen recalls, «worked with
Jan Hermans from Chapel Hill on the dynamics of water in a crystal of the protein
BPTI (*basic pancreatic trypsin inhibitor*), treating water with the new method of
constraint dynamics, and with Peter Rossky and Martin Karplus from Harvard on the
complete dynamics of a dipeptide in ST2 water». Let us explain first that dipeptides
are organic compounds of two amino acids. Amino acids are the building blocks
which, when bound in short chains, constitute molecules called peptides, whereas
when they are bound in long chains constitute macromolecules, among which there
are the proteins. As for ST2 water, this is a model for the simulation of water. We
shall soon discuss the new method of "constraint dynamics".

Hermann Berendsen recalls a few key points from the workshop. For instance,
John Ermak made a pioneering presentation on stochastic dynamics, aiming at
reducing, in the equations to be integrated, the number of variables related to the
water molecules keeping the protein in the solution, thereby reducing the amount of
calculations required without compromising the quality of the resolved model.

This was the stumbling block for those scientists coming to Orsay to study the
molecular dynamics of proteins. Awareness of the huge amount of calculations
involved is neither new, nor so old, after all. We could establish a significant date
and place where this awareness was acquired. The place was CECAM at Orsay, and
the date 1972. The main actor was Herman Berendsen, who organized the workshop
on the molecular dynamics of liquid water, as we said earlier. In that first meeting—
Berendsen recalls—the issue had been raised of how to apply molecular dynamics
to large biological molecules, such as proteins.[20] These biological macromolecules
generally act inside a water solution, in particular when they fulfil an enzymatic
function. Therefore, if you want to study them, you need to understand in detail the
nature of the inner interactions of the large protein, and their link with small, numerous molecules of liquid water. As a consequence, «the simulation of liquid water is
the first topic we've got to study».

However, in 1972, no one could predict whether molecular dynamics might be
applied to proteins. The study of these large molecules is much more difficult than
the study of liquid water only. Indeed, these molecules present a complex three-
dimensional structure with a peculiar folding behaviour in a water solution: it is this
structure that allows them to function like enzymes, and speed up chemical reactions
inside cells. The main difficulty lies in the fact that the vibrations of the molecule's

---

[19]This, and the following quotes from Berendsen, in H.J.C. Berendsen, *The Development of Molecular Dynamics at CECAM*, in J. P. Hansen, G. Ciccotti, H.J.C. Berendsen, *In Memoriam. Aneesur
Rahman. 1927–1987*, CECAM, Orsay 1987, pp. 9–12.

[20]H. Berendsen, *Introduction*, in *CECAM workshop "Models for Protein Dynamics"*, CECAM,
Orsay, 1976.

covalent bonds are very fast (of the order of $10^{-14}$ s), whereas the macromolecular folding inside a water solution, as Herman Berendsen recalls, is much slower, of the order of seconds, or even minutes. Treating these two phenomena together requires a huge computing power. It seemed impossible to meet this challenge, even with the most advanced computers available.

The 1976 workshop was prepared carefully and well in advance. It had been preceded in 1975 by a discussion meeting at CECAM, also organized by Berendsen, in order to arrange the workshop of the following year in the best possible way. The idea was to have a relatively small number of participants, but the expectations for the results were very high. And indeed the eight-week workshop between May and July did not disappoint those expectations.

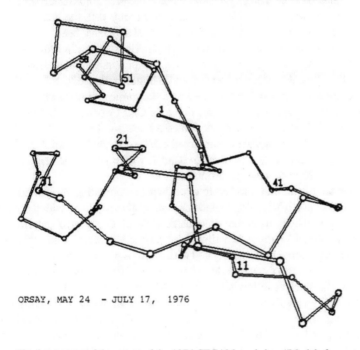

'MODELS FOR PROTEIN DYNAMICS'

ORSAY, MAY 24  – JULY 17,  1976

The front cover of the report of the 1976 CECAM workshop "Models for protein dynamics"

A number of pioneers in the studies of protein computer simulation were present. Shoshana Wodak, for instance, had completed an algorithm for the analysis of both protein-protein docking and the docking between a protein and other small organic molecules; Michael Levitt was developing a simplified representation of proteins; Peter Rossky and Andrew (Andy) McCammon, of the Karplus group at Harvard, and Wilfred Gunsteren, of the Berendsen group in Groningen, were engaged in specific studies of protein molecular dynamics.

This was the right time. Everybody arrived at Orsay with the latest news and a strong desire to obtain new results during the course of the workshop. Again, as Berendsen recalls, «during the workshop, Andy McCammon from Harvard—where the program had already been developed—realized the first simulation of the BPT1 protein in a vacuum, based upon the potential functions of proteins, worked out by Karplus'group». On the other hand, Berendsen goes on, «Van Gunsteren and myself changed the simulation program of proteins in order to incorporate all constraints».

The prospects for accelerating the progress of molecular dynamics were clear and had far-reaching implications. It was an historical turning point. Therefore, in order to provide a full account of it, we need to specify a few concepts mentioned by Berendsen. Let us start with the stochastic dynamic system, hinted at by the organizer of the workshop at Orsay in 1976. This is a dynamic system in which a part (for example, the irrelevant portion of the solvent/water) is replaced by stochastic collisions, as in Brownian motion. The physics to be represented is not affected; the irrelevant part, namely the water which does not interact directly with the macromolecule, is handled like noise. This is something which physicists have been doing for a long time, since it refers to those fluctuations which Albert Einstein himself studied in his papers on Brownian motion. In this way, the amount of calculation required to deal with the dynamics of the solvent *in toto* is hugely reduced, thanks to an approximation that can be handled in a much more economical way. It was therefore not by chance that this topic was discussed at the CECAM workshop devoted to models for protein simulation. Indeed, the action of a huge number of water molecules in which a large protein molecule is immerged can be considered "noise". John Ermak realized a numerical method in order to handle, with a stochastic representation of this kind, the portion of water far from the protein.

Another technique that allowed to cut down on computing needs and therefore make the simulation of complex molecules manageable had to do with the solution of the problem posed by the different time scales of the processes involved. In extremely complex systems, like those constituted by proteins in water solutions, there are several "degrees of freedom", associated with motions whose characteristic times can vary by several orders of magnitude. As we said before, this is a major problem for simulation, inasmuch as following all those different motions in detail would require extremely heavy calculations. A lucky coincidence would have it that, just before the beginning of the workshop at Orsay, Jean-Paul Ryckaert and Giovanni Ciccotti, tackling this problem, had perfected the constraint-dynamics method, which enabled scientists to make a qualitative leap, and to initiate in practical terms protein molecular dynamics. Berendsen was particularly interested in this method, and contributed to its development.

The history of the development of this method is, in broad terms, the following. In a workshop in 1974, Berendsen had launched the idea that some trick had to be found to reduce calculations. In the course of that workshop, André Bellemans, a Belgian physicist from the Université Libre of Brussels, specializing in polymer statistical mechanics and interested in molecular dynamics, decided that the young PhD student Jean-Paul Ryckaert should solve this problem; alone, since Bellemans abandoned his student at CECAM.

Ryckaert was at his wit's end, and talked about this with Giovanni Ciccotti. Thus a collaboration originated between them, which achieved a satisfactory conclusion in 1976, thanks also to the advice of Konrad Singer, and rare discussions with Berendsen, who still followed the matter with interest.

Ryckaert and Ciccotti solved the following problem: in systems of proteins in water solutions, there are numerous degrees of freedom, some of which can actually be frozen and therefore—de facto—removed. Such are, for instance, the inner vibrational movements of molecules, whose dynamics takes place at such high frequencies that they do not basically affect the properties interesting for the study of the behaviour of proteins, such as their folding in three-dimensional space, which take place in longer times. The solution of these vibrational motions would require a huge processing power, and therefore a huge amount of machine time. However, in the times in which these motions develop, the rest of the system does not change appreciably. Solving these motions not only engages huge amounts of calculation. It is also useless, because it prevents us from studying the really important motions of the molecule.

In other words, the detailed simulation of the dynamics of a macromolecule would require scientists to "follow" high-frequency vibrations of strong interatomic bonds and, as a consequence, would force them to follow time scales so small that it would be impossible to "see" the evolution of the structure as a whole, which usually takes place on time scales that are several orders of magnitude longer.

Ryckaert and Ciccotti found the way to get around this difficulty. The idea was to replace the fast vibrations of the strong covalent bonds with geometric constraints. This meant finding the right algorithm to solve the dynamics in Cartesian coordinates, while exactly maintaining the constraints throughout the course of the evolution. Once these fast motions had been removed, or rather, decoupled from the configurational motion of the molecules, one could use a sufficiently large integration step to reach the simulation times necessary to see the all-important (configurational) phenomena of the protein motion. In short, the use of both the constraints and the SHAKE algorithm, which enabled scientists to implement them, turned out to be an extraordinarily powerful tool for the simulation of proteins, and more generally, for the simulation of macromolecular systems.

Berendsen's and Singer's independent hints, which were useful for the fine-tuning of the algorithm, concerned diatomic molecules and algorithm implementation.

Let us go back to the period immediately preceding the workshop at Orsay. The SHAKE algorithm was completed in the spring of 1976: it had not been published yet when the CECAM workshop opened. Indeed, in those weeks, Ryckaert and Ciccotti were very busy drafting the paper they wanted to submit to the *Journal of Computational Physics*, the journal founded and directed by Berni Alder.[21]

The publication procedure started immediately after the conclusion of the workshop. Indeed, the *Journal of Computational Physics* received the paper on July 19,

---

[21] J.P. Ryckaert, G. Ciccotti, H.J.C, Berendsen, *Numerical integration of the Cartesian equation of motion of a system with constraints: molecular dynamics of N-alkanes*, Journal of Computational Physics **23**, 327, 1977.

1976, two days after the participants had left Orsay. It was signed not only by Jean-Paul Ryckaert and Giovanni Ciccotti, but also by Berendsen, who supplied an efficient implementation of the algorithm, thus realizing, together with Wilfred van Gunsteren, a smart computer program for solving the molecular dynamics of biological molecules.

In any case, even though SHAKE had not been published yet, it did give a remarkable boost to the workshop, since it provided a way to overcome the problem associated with the high frequencies of the chemical bond vibrations. Thanks to SHAKE, in the eight weeks of the workshop at Orsay in 1976, the molecular dynamics of proteins was actually born. It quickly exploded into a worldwide activity.

Indeed Berendsen, who knew the SHAKE algorithm by then, arrived at Orsay with his PhD student Wilfred van Gunsteren, who later became a professor at ETH Zurich. They had the specific assignment of including constraints in the molecular dynamics program they were developing to model macromolecules.

As to the two authors of the algorithm, Ryckaert and Ciccotti were not very interested in its application to biomolecular simulations, since they had solved the fundamental problem. In any case, the two scientists decided to take part in the workshop in order to find a smart solution to another problem, namely, the simplified handling of the solvent, by reducing it to a Brownian motion, in the wake of Ermak's procedure.

Be that as it may, the simulation of the molecular dynamics of the protein BPTI was introduced for the first time at the 1976 CECAM workshop at Orsay. This is why that eight-week workshop may be considered as the birth place of a new and very broad field of study, that of biomolecular simulation.

The higher computer processing power played a key role in this context. It was not just ordinary processing power, but rather the specific processing power at Orsay. Martin Karplus recalls: «During the 1976 workshop, Andy McCammon managed to realize the first simulation of a protein molecular dynamics, as detailed in his contribution in the CECAM report of 1976».[22] In that same year, Andrew (Andy) McCammon, after obtaining his PhD, had started his postdoc research work with Karplus at Harvard, in the USA. Karplus and his group were starting to use computer simulations to study the way proteins in solution change their shape. Andy immediately got involved in this study and, using a molecular dynamics software developed by Bruce Gelin, tried to estimate the potential energy of the conformation of several protein molecules. Together with Gelin himself and Karplus, Andy McCammon started to develop a molecular dynamics simulation of a specific protein, namely BPTI.

This was a difficult although not impossible enterprise to realize at Harvard. Referring to McCammon's report at the 1976 meeting, Martin Karplus comments: «What he does not mention is how hard we (he and Bruce Gelin, in particular) worked at Harvard to have the molecular dynamics program fully operational so that as soon as the workshop started he could begin the BPTI simulation on the Orsay CDC "supercomputer", an important element in making the extended workshops so

---

[22]M. Karplus, *Carl Moser and CECAM*, in the document mentioned in note 18 above.

successful.[23] It would have been difficult to do the 9.2 ps BPTI simulation in the USA, given the limited computational resources available to academic researchers at the time. Thus, Carl Moser, CECAM, and the Orsay Computer Center played a significant role in making possible the first molecular dynamics simulation of a protein, an essential element in the 2013 Nobel Prize in Chemistry, though not in the Nobel citation».

Martin Karplus receiving in Stockholm the 2013 Nobel Prize for chemistry, shared with Arieh Warshel and Michael Levitt. (Copyright @ Nobel Media AB 2013, Photo: Alexander Mahmoud)

Again, Berendsen recalls: «Although the application to biological molecules— such as proteins—had been developed in several places, Harvard in particular, by the group of Martin Karplus, it is probably fair to say that CECAM activities were inspirational in this field of study, and certainly stimulated further activities both in Europe and in the USA». In this statement, we may note Herman Berendsen's modesty, since he had also realized significant computer simulations of the behaviour of biological molecules.

Good ideas and computer power: this is the secret of molecular dynamics. In comparison with 1976, Emanuele Paci recalls forty years later, the second element— computer power—has grown dramatically. By now, molecular dynamics can deal with a very large number of biomolecules and longer times, of the order of hundreds of microseconds (one microsecond is equal to a millionth of a second). However, «the methodology has changed little since, and several of the present-day challenges

---

[23] Actually, in 1976, the supercomputer present in CIRCE was not a CDC but an IBM 370/168.

had already been clearly stated then».[24] The right ideas had already been put on the table at Orsay in 1976.

In sum, the workshop at CECAM between May and July was a source of great inspiration. Wilfred van Gunsteren recalls a climate of close and effective cooperation: «The participants that stayed for the whole two-month duration of the workshop did not only enjoy the goodies of Paris at the time, but also the close cooperation during eight weeks between scientists contributing different expertise to make progress towards the goal of simulating the behaviour of biomolecular systems. It goes without saying that Herman Berendsen was instrumental to this process of interaction».[25] It is also true that at Orsay, between May 24 and July 17, 1976, the SHAKE algorithm probably gave a significant boost to the study of the molecular dynamics of proteins and other biological macromolecules. Indeed, this algorithm would be used for many years to come, and it is still used today. Moreover, it triggered the imagination of several scientists specialized in applied mathematics, who were trying to identify the mathematical reasons for its validity. The successful features of this algorithm are the following: (1) it does not add to the error of any algorithm for the integration of the system motion; (2) it is "symplectic", i.e., it is one of the best algorithms for the integration of dynamics.

No doubt, as the workshop came to its end, all participants were convinced that a new window had been opened in the field of molecular dynamics. From then on, it would be possible to study the complex molecules of life.

---

[24]E. Paci, *Foreword*, note 12.

[25]W.F. van Gunsteren, *The Roots of Biomolecular Simulation*, in the document mentioned in note 18 above.

# Chapter 6
# Simulation Comes of Age

## 6.1 Effective Sampling: Charles Bennett and John Valleau

As we saw earlier on, once the soundness of the basic ideas had been checked, most results obtained from computer simulation in the first few years, both in Monte Carlo and in molecular dynamics, consisted in studying a few interesting kinds of behaviour (such as phase transitions) of ideal systems (hard spheres, rigid disks), or in the simulation of realistic models of more and more complex physical systems (from argon atoms to water molecules, down to the first attempts to reconstruct large biological molecules). The results were at the same time encouraging, and often temporary, and/or inconclusive. In particular, for an exhaustive study of the properties of more complex systems, or critical situations, such as the coexistence of different phases, or the onset of specific chemical reactions, it was soon clear that any advancement of simulation techniques would require the ability to control and calculate the thermodynamic quantities governing these processes. The relevant quantities are the thermodynamic potentials (entropy, enthalpy, free energy); the possibility of keeping them under control is a key factor in establishing the properties of specific processes. Starting from the mid-70s, the development of algorithms which allowed scientists to calculate thermodynamic potentials became one of the main lines of development of simulation.

To be fair, pioneering attempts in this direction had started even earlier on. Between 1967 and 1968, William (Bill) Hoover and Francis Ree, from Livermore Laboratory, published two articles in which they revisited the old problem of the phase transition within a hard-sphere system. They reinforced the results already acquired through the introduction of a calculation procedure which enabled them to evaluate the entropy of the system in its various stages.[1]

---

[1] W.G. Hoover, F.H. Ree, *Use of Computer Experiments to Locate the Melting Transition and Calculate the Entropy in the Solid Phase*, Journal of Chemical Physics **47**, 4873 (1967); W.G. Hoover, F.H. R, *Melting Transition and Communal Entropy for Hard Spheres*, Journal of Chemical Physics **49**, 3609, 1968.

© Springer Nature Switzerland AG 2020
G. Battimelli et al., *Computer Meets Theoretical Physics*, The Frontiers Collection,
https://doi.org/10.1007/978-3-030-39399-1_6

Hoover's interest in molecular dynamics dated back to 1962, when he obtained a PhD in physical chemistry at Michigan University, then went on to Livermore, attracted by the prospect of working with Berni Alder. Hoover was still a student when he convinced himself that molecular dynamics offered great opportunities; he was impressed in particular by the paper published by Alder and Wainwright in *Scientific American*, which we mentioned earlier on. Thus, as he had to choose between two possible jobs, the first at Los Alamos with Bill Wood, the second at Livermore with Alder and Wainwright, he chose the second. «Besides the better money for me it was much more interesting to see *motion* governed by differential equations rather than watching uncorrelated Monte Carlo *moves*, no matter how clever the underlying algorithm. Berni made it easy for me to learn and to work with him and some of his many colleagues, Brad Holian, Francis Ree, Tom Wainwright, and David Young. We worked on a variety of projects in kinetic theory and statistical mechanics, mostly directed toward equation of state properties for hard particles».[2]

Indeed, Hoover and Ree were also focusing upon the problem of hard spheres, and developed a method which allowed them to confirm what had already been proved by previous simulations, and accurately pinpoint where the fluid-solid transition took place by calculating the difference of entropy between the two phases. To this end, they had to perform a thermodynamic integration along a reversible path connecting the fluid to the solid phase. The problem was that, once the density of the system decreased, starting from the solid state, the transition took place suddenly in an irreversible way.

The difficulty was solved by devising a virtual hard-sphere system, where each sphere is "confined" inside its own cell in phase space; this assures the persistence of the solid phase even at low densities, and thus enables calculation of the entropy along a reversible path, passing through this idealized state. In the first of their two papers, Hoover and Ree illustrated the general idea, whereas in the second they showed the results, obtained by applying their "single occupancy" model to the hard-sphere system, while identifying the parameters of coexistence of the fluid and solid phases. Notwithstanding Hoover's preference for molecular dynamics, these papers used the Monte Carlo method.

In their conclusions, the two scientists from Livermore stated that «although we considered only hard particles in these calculations, there is no reason to expect any difficulty in extending these calculations, at high temperature at least, to more complicated systems». The extension was duly made shortly afterwards; not at Livermore, but rather at Orsay, by Loup Verlet and Jean-Pierre Hansen, who, in 1969, published a paper in which they showed the outcome of a simulation with the Monte Carlo method of the phase transitions of a system of particles interacting with a

---

[2]W.G. Hoover, *From Ann Arbor to Sheffield: Around the World in 80 Years. I.*; *Yokohama to Ruby Valley*, Computational Methods in Science and Technology **23**, 133, 2017.

Lennard-Jones potential.[3] The connection with the work made beforehand at Livermore was explicit in the abstract of the paper: «the fluid-solid transition has been investigated using a method recently introduced by Hoover and Ree».

In fact, Hansen and Verlet extended Hoover and Ree's approach, so as to cover the whole phase diagram of the system, evaluating the parameters of both the liquid-gas and the solid-liquid transition. For the liquid-gas transition they developed a technique similar to the "single occupancy cell" in order to force the system to remain homogeneous in the transition region. «Large density fluctuations leading to a gradual phase separation are thus prevented and a reversible path joining the gas to the liquid phase can be constructed [...] Integration of the pressure along this continuous isotherm yields the liquid-phase free energy and this in turn allows the determination of the transition data». This procedure was repeated for the fluid-solid transition, using the method developed by Hoover and Ree. This was the first paper in which, through computer simulation, scientists managed to calculate the free energy of a system of particles interacting with a Lennard-Jones potential, thus locating the transitions and reconstructing the whole phase diagram.

These papers were published in the pioneering stage of simulation. Rahman's first papers about real atomic systems had appeared a few years earlier, and the simulation of polyatomic molecules, such as water, was still to come. However, in the early 70s, the situation changed significantly, as it became clear that increased computing power and the appearance of new algorithms would make it possible to study much more complex systems.

In the acknowledgements at the end of their 1969 paper, Hansen and Verlet mentioned, beside the contribution of Dominique Levesque, also of the Orsay group, the critical reading of the manuscript by John Valleau. Valleau was a chemist from Toronto, where he had studied, whereas he later obtained a PhD in England, where he worked at Cambridge under the supervision of Christopher Longuet-Higgins and subsequently studied several problems in both theoretical and experimental chemistry. In 1961, he was appointed professor of theoretical chemistry at the University of Toronto, where there was already a strong experimental chemistry group led by John Polanyi, and started to build a theoretical chemistry group, which would soon become the largest in Canada.

---

[3]J. Hansen, L. Verlet, *Phase Transitions of the Lennard-Jones System*, Physical Review **184**, 151, 1969.

Ray Kapral and John Valleau, early 2000s (R. Kapral)

In 1968, Valleau spent a sabbatical year at Orsay, where he came into contact with Verlet's group and familiarized himself with simulation techniques. Back in Toronto, he started getting directly interested in the problem of free energy computation, looking for methods to obtain accurate Monte Carlo estimates by using more efficient algorithms. In a series of papers, published at first in collaboration with Damon Card, and later on with his PhD student Glenn Torrie, he gradually refined the idea of increasing the efficiency of calculation by sampling primarily those states which were closer to those of interest, instead of considering the whole range of configurations.[4]

In statistical mechanics, free energy is directly linked to the probability of a state, which, in turn, is measured by a term (the "Boltzmann factor") which "weights" each particular configuration. If one is interested in regions characterized by a low probability, this procedure gets extremely expensive and inefficient with the Monte Carlo method, since random sampling tends to select highly probable states, and rarely visits further, possibly interesting though less probable states. The idea is to force sampling to preferably select the improbable states, associated with "rare events", thus redefining the probability distributions, so as to assign a higher weight to these events, and then re-scaling the resulting histograms to obtain the right estimates of the average values, which provide the calculation of the thermodynamic quantity (in this

---

[4] J.P. Valleau, D.N. Card, *Monte Carlo Estimation of the Free Energy by Multistage Sampling*, Journal of Chemical Physics **57**, 5457, 1972; G. Torrie, J.P. Valleau, A. Bain, *Monte Carlo estimation of communal entropy*, Journal of Chemical Physics **58**, 5479, 1973; G. Torrie, J.P. Valleau, *Monte Carlo Free Energy Estimates Using Non-Boltzmann Sampling: Application to the Sub-Critical Lennard-Jones Fluid*, Chemical Physics Letters **28**, 578, 1974.

part of the algorithm, they resurrected the technique of "histogram re-weighting", developed in 1967 by McDonald and Singer).

In those same years, Charles Bennett was working on the same problem on his own at the IBM Research Centre in Yorktown Heights, New York State. Bennett had also been trained as a chemist; directed by David Turnbull, he started working—for his PhD dissertation—on the random packing modes of hard-sphere systems, using the IBM7094, which had just been bought by the new computer science centre in Harvard. Turnbull soon had the idea of getting in touch with Berni Alder («who represented the state-of-the-art in computer dynamic simulation of both disks and spheres»), and Bennett ended up spending a few summers in Berkeley, interacting with Alder and Mary Ann Mansigh at Livermore, and in 1971 completing a dissertation in molecular dynamics, calculating the gaps in crystallized hard-sphere systems. In the same year, he published with Alder the ninth paper of the series of *Studies in Molecular Dynamics*,[5] directly linked with the last section of his dissertation. For some time, he continued studying molecular dynamics, working with Rahman at the Argonne laboratory, but he soon moved on to the IBM research division and gradually abandoned the field of simulation; he started working on quantum cryptography and quantum information theory, and became one of the founders of the research on teleportation and quantum entanglement. Before that, however, he left a significant mark in the progress of simulation.

One of the main problems, which make simulation either too expensive or inefficient, is the coexistence, in complex systems, of motions characterized by very different characteristic frequencies: the task of accurately following the high-frequency motions requires a huge investment of machine time in order to handle low-frequencies ones as well. Inspired by the discussions with Rahman, who was a specialist in developing unorthodox ways to make molecular dynamics methods more efficient, Bennett invented a technique which enabled him to slow down the high-frequency motions and speed up the low-frequency ones, thus increasing the effectiveness with which simulation could explore the configuration space in a given time.

The idea was to change the expression of the kinetic energy of a system of particles by introducing, instead of real masses, fictitious ones (more precisely, one should replace the kinetic energy with a more general quadratic function of velocity, in which a "mass tensor" appears); if the new masses are properly chosen, they can produce the desired effect of "homogenizing" the frequencies of the various motions characterizing the system and, although they produce different dynamics from the actual one, they lead to the same configurational equilibrium properties of the original system. This method was explained in a paper, published at the beginning of 1975.[6] It did work: the test made upon a short polymer chain (ten atoms) with a Lennard-Jones potential showed that the method had saved machine time by a factor between 5 and

---

[5]C.H. Bennett, B.J. Alder, *Studies in Molecular Dynamics IX. Vacancies in Hard Sphere Crystals*, Journal of Chemical Physics **54**, 4769, 1971.

[6]C.H. Bennett, *Mass Tensor Molecular Dynamics*, Journal of Computational Physics **19**, 267, 1975.

10. Bennett underlined that «a more dramatic improvement could be expected in a system having a greater disparity between its hard and soft modes».

However, one could still not observe the changes in shape typical of a polymer chain. This led Bennett to a conclusion which opened the door to research in a complementary direction: «The failure of the 10-atom chain to undergo a major conformational transition, even with the assistance of mass tensor dynamics, emphasizes the distinction between events that are infrequent primarily because they involve soft, slowly relaxing degrees of freedom and events whose infrequency results primarily from the presence of a large activation barrier. Events of the latter type need not be associated with any particularly soft modes of the force constant matrix, and may remain quite infrequent under mass tensor dynamics. Such events are probably best studied by molecular dynamics and Monte Carlo methods in which an artificial constraint force is used to push the system into the relevant saddle point neighbourhood in configuration space». In sum, Bennett suggested, in order to study "rare events", one should develop either molecular dynamics methods which force the system to go where one would like to study it, or Monte Carlo techniques that concentrate sampling upon improbable regions: the latter was the research line which John Valleau was following at the time in Toronto.

The contributions of Valleau and his team were explicitly mentioned in the paper which Bennett published the following year, where he discussed the general problem of free energy estimation, and developed a method to optimize calculation in Monte Carlo simulations.[7] Bennett pointed out that, as a general rule, calculating free energy required a procedure, similar to calorimetry, enabling to connect the system of interest through a reversible path with a reference system of known properties. This "computer calorimetry" offered the great advantage that the reference system (or the intermediate systems considered along the way) could be chosen arbitrarily with regard to the system under study, thus changing not only the thermodynamic variables of state, but also the potential, which may be useful to satisfy specific conditions. In the first work done at Livermore and Orsay, for example, an additional constraint was built to stabilize the system against phase transitions, whereas in the simulations realized in Toronto, additional terms were introduced to concentrate the probability density in specific portions of the configuration space.

Bennett faced the general statistical problem of establishing how, whatever the procedure and the system used as a reference, one might draw—from simulation data—the best estimate of the free energy difference between the systems being considered. He managed to do this by developing a method using what he called the *acceptance ratio*. In this, he showed that the problem may be reduced to calculating the ratios of the integrals of the probability distributions, which could be inferred from the data gathered in the simulation of the two systems in the same configuration space, but with different potentials; the efficiency of the method increases with the degree of overlap between the two sampled sets. Bennett also showed that one could obtain a good estimate of the free energy even in the presence of a small overlap, by

---

[7]C.H. Bennett, *Efficient Estimation of Free Energy Differences from Monte Carlo Data*, Journal of Computational Physics **22**, 245, 1976.

using a technique of data interpolation, which was justified if the density of layers in each set was a "smooth" function of the difference of potential. This condition was often satisfied in the physical systems being studied.

In his conclusions, Bennett stressed that some aspects of his method could be traced back to the 1972 paper by Valleau and Card, recognized as «apparently the first calculation of free energy with overlapping techniques», but deemed as «a rather complicated procedure [...] with a definitely lower statistical efficiency», and mentioned, as examples of a later improved technique, the papers published by Torrie and Valleau in 1973 and 1974, and another one, in the process of being published. Indeed, that latter was published a few months after Bennett's own paper, in the journal founded by Alder, the *Journal of Computational Physics*, at the beginning of 1977.[8]

Torrie and Valleau remarked that the standard technique for calculating free energy (numerical integration over a series of points connecting the system under study to another one with known energy), beside being at the same time inelegant and heavy, was also not efficient enough, although not completely useless, in phase transitions, because of the difficulty in defining an integration path along which to measure the necessary averages in a reliable way. On the other hand, it was in just these situations that the free energy estimate would be most useful. Torrie and Valleau pointed out that they had already obtained promising results for the liquid-gas transition of a Lennard-Jones fluid using Monte Carlo sampling on an arbitrary distribution, constructed so as to explore those parts of the configuration space common to the fluid and the reference system, and defining appropriate re-calibrations of the data to get the correct average values of the thermodynamic variables. The two scientists could now show the possibility of extending the techniques developed in that case to explore systematically large regions of the phase diagram.

This new technique was referred to in the title of the paper itself as *umbrella sampling*. «To facilitate having the method noticed and adopted by the simulation community, it needed a name. I advocated for "non-Boltzmann sampling" on the grounds that the core of the approach was the recognition that the Boltzmann weighting of states that emerges from ensemble theory was not the most efficient way to drive a Markov chain exploration of the relevant regions of configuration space in a computer experiment. John had misgivings about this and countered with *umbrella sampling*. When I agreed to this, it was under the impression that the choice was inspired by the visual appearance of the broad, flat energy distributions that were the goal of the method [...] It finally came out in conversation that his actual intent had been to emphasize the versatility of the approach in covering many different possibilities».[9]

---

[8] G.M. Torrie and J.P. Valleau, *Nonphysical Sampling Distributions in Monte Carlo Free-Energy Estimation: Umbrella Sampling*, Journal of Computational Physics **23**, 187, 1977.

[9] Letter from G.M. Torrie to G. Ciccotti, February 2019.

## 6.2   Molecular Dynamics for Generalized Ensembles

We have seen that a direction successfully followed by molecular simulation in the
first twenty years from its birth was its ability to handle more and more realistic
models, suitable to represent more and more complex systems: from hard spheres to
argon and water, to the problem of macromolecules and proteins. A complementary
direction, which witnessed significant steps forward in the next few years, consisted
in realizing simulations in which macroscopic conditions other than energy, volume,
and number of particles (E,V,N) were kept under control during the system evolution.
In the language of statistical mechanics, this meant moving the phase space point of
the system in such a way that the points asymptotically visited were sampling various
ensembles, not only the original (E,V,N); the extension of molecular dynamics to
different ensembles was an expression of the explosion of creativity, typical of the
mature phase of molecular simulation.

The concept of ensemble was proposed for the first time in 1902 by Josiah Willard
Gibbs, and consists in specifying the probability distribution of the microscopic states
of a system for which fixed values of some thermodynamic quantities are assumed.
In other words an ensemble can also be the (infinite) population of microscopic
states distributed according to its probability distribution (in probabilistic terms, a
full realization of the random variable "phase space points"). For example, if we
consider a system with a number N of identical elements staying within a given
volume V, submitted to the condition that its total energy E be constant, the representative ensemble of the microscopic states accessible to the system is called the
*microcanonical ensemble* in statistical mechanics. If, on the other hand, we wish to
keep the temperature T fixed, rather than the energy, we pass from the microcanonical
(N,V,E) to the *canonical ensemble* (N,V,T).

Now, the Metropolis Monte Carlo method can easily be used to produce asymptotically any ensemble with given probability distributions, because it is not difficult
to define a "Markov chain" suitable to do the job. The "Markov chain", in turn, is
a peculiar stochastic process, in which the probability that the system assumes a
certain (microscopic) state depends only on the immediately preceding state, rather
than the path leading to that state. The latter property, a sort of "short-term memory",
is referred to as the "Markov property". In particular, as in the first paper published
by Metropolis and colleagues, the simulation with the Monte Carlo method asymptotically sampled a canonical ensemble, i.e., an ensemble corresponding to a given
number of molecules (N) and fixed values of the volume (V) and temperature (T).

The situation is not as simple as that in the case of molecular dynamics. In the
first simulations of molecular dynamics, one studied the behaviour of a fixed number
of "primitive" elements (hard spheres, argon atoms, water molecules), confined in a
definite volume, and with a fixed value of the total energy. The values of the physical
properties of interest were calculated using suitable averaging operations on the
states successively occupied by the system in the course of its asymptotic evolution
in time. At the start, molecular dynamics was therefore sampling the microcanonical
ensemble, where, apart from the number of particles (N) and the volume (V), the

value of the energy (E) is also fixed. This was a natural, actually mandatory choice, since energy is the quantity conserved by Newton's laws of motion. However, it might be desirable for the simulation of realistic physical processes to let the system evolve under different conditions, for example by keeping the value of the temperature (T) fixed rather than the energy, since this is a parameter which is easier to control experimentally; otherwise, one could let the volume vary, while keeping the pressure (p) fixed, as actually happens in the case of most phase transitions from the liquid to the solid state.

The possible extension of the techniques of molecular dynamics to sample from an equilibrium ensemble other than microcanonical was opened up in 1979, when the US scientist Hans Christian Andersen realized a molecular dynamics scheme for systems with constant pressure and enthalpy, by introducing new continuous dynamics, close to but definitely different from the "true" Newtonian dynamics, in which the volume became a new degree of freedom whose motion was coupled to that of the particles. To derive the equations of motion of this new system, Andersen constructed a suitable new Lagrangian. He could demonstrate that this new dynamics could bring a system to equilibrium. Moreover, the subsequent trajectory sampled the equilibrium distribution of a less common but perfectly legitimate ensemble in which the number of particles, pressure, and enthalpy were fixed. To recover the most useful condition of an ensemble with given N, V, and T, Andersen added another ad hoc discrete procedure which he could demonstrate would sample the standard canonical ensemble under stationary conditions.[10] Of course, in both cases the new dynamics introduced would be in principle unphysical, hence no good to study time-dependent properties. However, if the induced perturbation of the dynamics was small enough, one could hope that even time-dependent properties could be computed. In practice, this condition is indeed often satisfied but no mathematical proof is available to say that this is true in general, at least for sufficiently weak volume-particle coupling.

Hans Christian Andersen was a physical chemist, born in 1941 in Brooklyn, New York. He obtained a PhD at the Massachusetts Institute of Technology (MIT) in Boston, where he acquired the knowledge needed to apply a mix of pure mathematical techniques of statistical mechanics to the solution of chemical and physical problems. In 1968 he started work as assistant professor at the Chemistry Department of Stanford University, California. In 1974, he was promoted to associate professor of chemistry, and then to professor of chemistry in 1980, the year of publication of his celebrated paper.

Around 1979 an idea was developed which opened the door to a new field of application of computer simulation. In that period, Andersen was focusing, among other things, on supercooled liquids and amorphous solids. He realized that molecular dynamics might help him, but also that the available simulation techniques were not well suited to study his particular problem. He had to invent something new. He remembers: «I wanted to be able to simulate the behaviour of a liquid when it is cooled at a steady rate to temperatures well below its freezing point. It was

---

[10]H.C. Andersen, *Molecular dynamics simulations at constant pressure and/or temperature*, Journal of Chemical Physics **72**, 2384, 1980.

straightforward to do this if the volume of the material were held fixed. But real glass forming materials contract as the temperature is lowered at constant pressure. I suspected that this change in density, even if it is rather small, might have an important effect on the relaxation rates of a supercooled liquid».[11]

Andersen did not only want to simulate the phenomenon, he was also particularly interested in studying how the rate of cooling of the liquid affected the structure and relaxation rates of the supercooled liquid; that is why he had to rely on the simulation techniques of molecular dynamics. The US physical chemist certainly knew that, with the Monte Carlo method, he could make simulations of the system at constant pressure, with a temperature which was varied in a controlled way. «However, I had no reason to think that the dynamics generated by Monte Carlo transition probabilities would provide a realistic description of energy conserving Hamiltonian dynamics».

Forty years later, Andersen did not remember precisely how ideas developed in his mind. The method he obtained resulted from the combination of Lagrangian mechanics (the fundamental procedure from which to derive equations of motion of any general mechanical system) with the original idea of introducing the volume as an "additional coordinate" together with the usual particle coordinates. A coordinate transformation was devised so that the transformed particle coordinates would obey simple periodic boundary conditions even though the volume was changing with time. In this way, he introduced a new mechanical system in which the pressure and a mechanical quantity connected to the thermodynamic variable enthalpy were held fixed, thus managing to establish a direct connection between mechanics and the traditional concepts of statistical thermodynamics. Moreover, given that the new system did not fix the system temperature, Andersen managed to «obtain a method of molecular dynamics at constant temperature and pressure» by starting from a isobaric-isoenthalpic ensemble (at constant pressure and enthalpy) and mixing his dynamic approach and an appropriate algorithm that chooses a particle at random from time to time and assigns to it a new velocity chosen at random from a Maxwellian distribution corresponding to the given temperature.

Three months later, Andersen submitted his paper to the *Journal of Chemical Physics*. He wrote: «In the molecular dynamics simulation method for fluids the equations of motion for a collection of particles in a fixed volume are solved numerically. The energy, volume, and number of particles are constant for a particular simulation, and it is assumed that time averages of properties of the simulated fluid are equal to microcanonical ensemble averages of the same properties. In some situations, it is desirable to perform simulations of a fluid for particular values of temperature and/or pressure or under conditions in which the energy and volume of the fluid can fluctuate. This paper proposes and discusses three methods for performing molecular dynamics simulations under conditions of constant temperature and/or pressure, rather than constant energy and volume».[12]

---

[11] This and the two following quotes come from a letter from H.C. Andersen to G. Ciccotti, December 2018.

[12] H.C. Andersen, *Molecular dynamics simulations at constant pressure and/or temperature*, cited in note 10.

This was an outstanding result which provided a way to sample, by a continuous dynamical trajectory, microscopic states distributed according to a given ensemble, no longer necessarily microcanonical. These states could then be used as starting states for physical, dynamical simulations. Moreover, the new dynamics was shown to be able to drive an arbitrary initial state to the equilibrium states of the given ensemble. As we will see shortly, these were the things that influenced Rahman/Parrinello and Nosé to introduce additional coordinates to produce useful (pseudo-)dynamics able to sample the best ensemble for the problem under scrutiny.

Ironically enough, this splendid result (for equilibrium simulations) was not Andersen's original motivation. He wanted to simulate the rapid cooling of a liquid. If one tries to produce an amorphous state from a fluid with a procedure at constant volume, one misses all the experimental physics measuring first and foremost, at constant pressure, the reduction in volume of the substance with a lowered temperature (in other words, the growth in density of the solid state—whether it be amorphous or crystal—which is produced by solidifying a fluid). There was, therefore, a remarkable limit to the power of simulation. This was the limit which Andersen wanted to remove. However, his result did not exactly follow the direction that initially motivated him: «The methods discussed above», he wrote, «were not ultimately useful for understanding the dynamics of supercooled liquids near the glass transition. At the time I did not realize that the phenomena that take place in supercooled liquids near the laboratory glass transition, which happen on very long time scales, are distinctly different from what can be observed in simulations, whose duration is only hundreds or thousands of picoseconds».[13]

Andersen developed his method in splendid isolation, without the help of any colleague or student. Evaluating the meaning of his work later on he estimated that he got "a new appreciation of the fact that computer simulations are not just methods for investigating the mechanics of atoms and molecules (i.e., the way they move). They are a tool for studying the classical statistical mechanics and statistical thermodynamics of matter". Above all, by enlarging this theoretical perspective of simulation, he thought that the importance of his contribution consisted in the fact that it «encouraged the development of yet more new simulation techniques and provided some tools to implement those developments».

The first of these developments was made by Anees Rahman and Michele Parrinello. They extended the method developed by Andersen to handle the variations in volume of the simulation cell and hence even to handle the variations in the shape of the cell, and be able to define the crystal structure implied by the various empirical additive pair potentials in use at the time. The initial idea was that the fixed periodic boundary conditions used in conventional molecular dynamics simulations exclude the direct observation of solid-solid phase transitions because the boundary conditions, chosen to be compatible with a solid phase, are not in general compatible with the others. As a consequence, fixed periodic boundary conditions tend to stabilize a solid phase well beyond its regime of thermodynamic stability, and can easily lead one to completely overlook the existence of other, more stable phases. The solution

---

[13]This quote and the following are taken from the letter H.C. Andersen to G. Ciccotti, see note 11.

consists in allowing the shape of the periodic box in which the particles are confined to change.

We already know about Anees Rahman being one of the founders of molecular dynamics. Michele Parrinello was a young Italian physicist, born in Messina in 1945. He obtained a degree at Bologna University in 1968 with a dissertation on the quantum-relativistic theory of fields applied to particle physics. He wrote: «All the best people were working in that field at the time, and as I was a good student I also worked on it».[14] Later on, Parrinello met Mario Tosi, a leader in condensed matter physics, who convinced him to change his interests from particle physics to statistical mechanics. «Tosi was my teacher, and I am indebted to him for this change of subject, which was in the end very fortunate», Parrinello remembered. Thus, the young physicist first went to Messina, where Tosi used to teach, and later on moved to Trieste with his master. And so it was that, in 1977, Michele Parrinello, aged 32, became assistant at Trieste University. Three years later, he took a sabbatical and flew to Argonne in the USA. There the Italian physicist, by now an expert in statistical mechanics, met Rahman and molecular dynamics.

Andersen's papers had recently been published, and Rahman was particularly excited at the idea of the possible developments. He was brooding over a related idea. Thus he could not resist the temptation to try to convince a theoretical physicist—whom he had met by chance, and considered capable of formulating a problem in a mathematical way—to help him realize a generalization which he judged very important. The question was: why not extend Andersen's approach to crystalline solids, and try to predict the structures of the elementary crystal cell implied by the various empirical potentials used in simulations? The answer to this question was not so simple. Rahman had experience in computer simulations. On the other hand, Parrinello was well versed in statistical mechanics. The two scientists complemented each other. Rahman found a way to convince Parrinello to change his scientific career once again. The two scientists started working, each exploiting his own respective skills, so that they soon managed to extend Andersen's approach to the simulation of phase transitions in crystals. The result was a specific application of Andersen's fundamental work, which allowed them to vary not only the volume, but also the shape of the crystallizing system. The paper by Rahman and Parrinello was published in 1980 in *Physical Review Letters*, and was entitled *Crystal structure and pair potentials: A molecular-dynamics study*.[15]

---

[14]D. MacKernan, *Feature Interview, Michele Parrinello*, in: SIMU. Challenges in Molecular Simulations, Newsletter, Issue 2, November 2000, p. 7; https://www.researchgate.net/publication/267979729_SIMU_Challenges_in_Molecular_Simulations_Bridging_the_Length-_and_Timescales_gap_Volume_2.

[15]M. Parrinello, A. Rahman, *Crystal Structure and Pair Potentials: A Molecular-Dynamics Study*, Physical Review Letters **45**, 1196, 1980.

«With use of a Lagrangian», the two wrote, «which allows for the variation of the shape and size of the periodically repeating molecular-dynamics cell, it is shown that different pair potentials lead to different crystal structures». The basic idea and procedure were not substantially different from Andersen's. However, Rahman realized at once that the same procedure, applied to crystals—anisotropic materials— would also provide a way to see the structure that the various empirical potentials used in simulations would lead to, thus going one step forward in the realism of molecular dynamics. In other words, the basic idea belonged to Andersen, while Rahman and Parrinello applied it also to crystalline solids, not just amorphous solids. Indeed, crystalline solids can have anisotropic cells, so it is not enough to take into account only the variation in volume: the shape of the crystallizing system also plays a fundamental role.

The breakthrough by Rahman and Parrinello was therefore not so much a conceptual innovation. Rather, it was an extension of the domain of physical reality accessible to the techniques of molecular dynamics. These could now also simulate crystallization processes. At this point, it became possible to make computer simulations of liquid-solid transitions, which are slow processes, therefore difficult to follow with simulation, unlike solid-liquid transitions, which take place quickly and do not require special expedients.

A further important step forward in extending the methods of molecular dynamics took place soon after, thanks to a Japanese theoretical chemist. In 1983, Shuichi Nosé submitted two papers on molecular dynamics simulation of canonical ensembles to two different scientific journals. These are ensembles in which the temperature is held constant. The algorithm he proposed would soon be known as the "Nosé thermostat". The two papers were published with a certain delay, in 1984, due to the fact that the referees found it difficult to accept such a new and original proposal.[16] His first paper, *A molecular-dynamics method for simulations in the canonical ensemble*, was published in *Molecular Physics*, immediately followed by the second, *A unified formulation of the constant temperature molecular-dynamics methods*, in the *Journal of Chemical Physics*.[17]

---

[16]Y. Kataoka, M.L. Klein, *Shuichi Nosé*, Physics Today, February 2006, pp. 67–68.

[17]S. Nosé, *A molecular-dynamics method for simulations in the canonical ensemble*, Molecular Physics **52**, 255, 1984; S. Nosé, *A unified formulation of the constant temperature molecular-dynamics methods*, Journal of Chemical Physics **81**, 511, 1984.

Schuichi Nosè (1951–2005) (Physics Today)

Shuichi Nosé was born on June 17, 1951 in Kyotango, near Kyoto. He had a background as a chemist. After getting his first degree at the university of Kyoto in 1979, Nosé continued his studies at the department of chemistry of Kyoto University, where he worked with Tsunenobu Yamamoto on problems of nuclear physics, applied to a methane molecule. While he was working on his PhD dissertation, Nosé started using molecular simulation, applying it to the classical model of phase transitions of solid methane, initially proposed by Hubert James and Thomas Keenan.

Once he obtained his PhD, in 1981, Nosé moved to Canada, first, for a short time, at the Institute for Material Research of McMaster University, Hamilton, Ontario, and later on in Ottawa, as a fellow of the Natural Sciences and Engineering Research Council, which, in turn, is part of the chemistry division of the National Research Council Canada (NRCC). In 1984 he became associate researcher at NRCC.

Shuichi Nosé stayed in Canada for three years. In this period, his scientific creativity really took off, accompanied by a high level of productivity. Indeed, he published several papers. In particular, as Yosuke Kataoka and Michael Klein wrote, Nosé «contributed to the methodological extension of the schemes of Molecular Dynamics by Andersen-Parrinello-Rahman, which make it possible to simulate on a computer the structural phase-transitions of molecular crystals». During these three years, Nosé developed his greatest contribution to the development of computer simulation: the "thermostat" which would take his name.

The two papers submitted by Shuichi Nosé presented an algorithm studying the behaviour of a system in contact with an energy reservoir (the thermostat). The reservoir keeps the temperature constant during the dynamic evolution. In principle, simulating the thermostat means immersing the physical system being studied in another, much larger system, which stabilizes the temperature. A direct simulation of the thermostat would be extremely expensive. The innovation of the young Japanese scientist consisted in reducing the huge number of degrees of freedom of the thermostat to only one. Doing this meant that the amount of computing necessary for simulation of the whole system was drastically reduced, and the thermostat could thus be "handled" computationally, assuring an unchanging temperature.

This was a very important result. Phase transitions take place at a fixed temperature; therefore, the availability of a computing algorithm allowing one to keep this variable under control with a continuous time-reversible dynamics was a remarkable step forward in comparison with the more cumbersome techniques used in the earlier molecular dynamics simulations already discussed.

Thus molecular dynamics in the canonical ensemble was fully realized for the first time. In "classical" molecular dynamics, simulations were realized in the microcanonical ensemble, that is, in systems with fixed particle number, volume, and energy. On the other hand, in experiments with real systems, temperature, rather than energy, is kept constant. The most suitable representative ensemble for the study of such systems is therefore the canonical ensemble. We have also seen that Andersen was the first scientist to suggest a technique for the study of the molecular dynamics of systems at constant pressure and temperature. However, his was a "mixed" algorithm—partly dynamic, partly stochastic. On the other hand, Nosé elaborated a direct algorithm which made it possible to fix the temperature of the system one hoped to simulate, as well as the fluctuations of the kinetic energy around the average, typical of the distribution of the canonical ensemble. That was a very clear step forward, which warranted immediate acknowledgement for its author, once the referees' doubts had been overcome.

The original model of Nosé's thermostat, however, had one limitation: the simulation it enabled was not easy to visualize, hence not so practical for the interpretation of results. This difficulty was overcome by Bill Hoover, who in a few months reformulated Nosé's equations of motion in terms which made the algorithm more intuitive and more useful for the interpretation of the results.[18] As a consequence, the fine-tuned technique was subsequently referred to as the "Nosé-Hoover thermostat".

In this case, unexpected coincidences helped. In the month of August 1984, Hoover was in a train station in Paris. He had just read the two papers by Shuichi Nosé and had been very impressed by them. Suddenly, on a platform, he saw a suitcase with the surname NOSE on it. He raised his head: the suitcase owner must be

---

[18]W.G. *Hoover, Canonical dynamics: Equilibrium phase-space distributions, Physical Review A* **31**, *1695, 1985.*

Japanese. He worked up courage, drew closer and asked: are you, by any chance, that Shuichi Nosé? And he actually got the answer he was looking for: yes. Chance would have it that the US scientist from Livermore Laboratory and his Japanese colleague, temporarily posted in Canada, had actually met in Paris, to participate, a few days later, in the same workshop on computer simulation, organized at CECAM, and got to know each other at the station. This chance meeting immediately developed into a friendship and, since they were staying in nearby hotels, close to Notre Dame, they had plenty of free time for passionate discussions about molecular dynamics and Nosé's papers.

«I learned enough from our Paris conversations», Hoover recalled, «to write a paper shortly thereafter (while visiting Philippe Choquard in Lausanne), stressing the importance of what is called the "Nosé-Hoover" version of his dynamical equations».[19] His paper was published in the month of March 1985.

Bill Hoover

At the end of 1984, Nosé went back to Japan. He had been called back by Ryogo Kubo, who offered him a position at the Physics Department of Keio University in Minato, Tokyo. Shortly before this, as we just mentioned, Nosé took part in the

[19]W.G. Hoover, *Nosé Shuichi, 17 June 1951–17 August 2005, In Memoriam*, http://www.williamhoover.info/nose.pdf.

CECAM Orsay workshop on *Constraint techniques in the simulation of transport and structural phase transitions*, where he met Hoover. Among the invited speakers of the workshop there was Andersen, who—unlike Hoover—had still not read Nosé's papers: «I was unaware of his work when I got the invitation to attend the CECAM workshop—Andersen recalled—, which was a wonderful experience for me».[20]

Once again, CECAM played a crucial role in those years, as the main "centre for the exchange of ideas" in molecular simulation. It was indeed the place where more and more scientists—specialized in different topics and coming from research institutes all over the world—could meet and talk, thus developing new ideas on computer simulation, by now a subject matter of primary importance for the scientific community.

Group picture of the participants to the CECAM workshop on "Solid State Diffusion", August 1983. From the left, lower row: V. Rosato, G. Ciccotti, A. Ladd, A. Rahman and W. Hoover; standing, S. Yip, C. Moser, R. Harrison, V. Pontikis, G. Kalonji and M. Guillopé

According to Andersen, who had worked all by himself on his own method, the workshop at Orsay was also "a wonderful experience" because it set him directly in touch with some of those who had realized the relevance of his method and had developed it further: «I remember being delighted when I received a preprint from Rahman and Parrinello on their simulation method in which the shape of the sample could vary. I remember being surprised and amazed at what Nosé did a few years later to apply my constant pressure idea to temperature».

---

[20]This quote and the following one come from a letter from H.C. Andersen to G. Ciccotti, dated December 2018.

At the end of the workshop, while Hoover started working, stimulated by his conversations with Nosé, to produce the modified version of the technique which would be published shortly afterwards, Andersen and Nosé were among the members of a restricted group of participants who were asked to prepare a report about *"New molecular dynamics methods for various ensembles"*. The workshop had not been organized on this particular aspect, but this topic emerged thanks to its own strength and relevance, as well as to the need, which came up during the Orsay workshop, to assess the situation in the field, since «the range of methods at constant temperature and pressure, the relationships among them, and their connections with the principles, generally accepted in Statistical Mechanics, caused a considerable confusion».[21] The report was contained in the CECAM activity report of 1984.

**RAHMAN FESTSCHRIFT**
**November 12-13, 1984**
**Argonne National Laboratory**

Argonne Laboratory, Rahman Festschrift, 1984: almost all the protagonists of our story can be found in the group picture. The meeting celebrated the 20th anniversary of the publication of Rahman's paper where for the first time the molecular dynamics simulation of a realistic model for Argon was realized (courtesy Argonne National Laboratory)

---

[21]H.C. Andersen et al., *New Molecular Dynamics Methods for Various Ensembles*, CECAM Workshop (Orsay, August 20 - September 1, 1984), CECAM, Orsay 1984, p. 86.

## 6.3   Liquid Crystals, Entropy, and Disorder: Daan Frenkel

The organizer of the CECAM workshop in the summer of 1984 was Daan Frenkel, yet another experimental chemist who was attracted by simulation. Frenkel was born in Amsterdam in 1948. He was the second of the four children of Maurits Frenkel and Herta Tietz, both medical doctors, who had escaped deportation during the Second World War. Daan studied chemistry, but had a deep interest in physics. He obtained his Master's degree in physical chemistry in 1972. In that same year, he published his first paper modelling the effect of collisions on rotation in dense liquids, at that time an active topic of research in physical chemistry. However, as Daan later commented, «that work taught me that collision models were doomed, as collisions in dense media are ill-defined. That is why I moved to Molecular Dynamics».[22]

Daan's first direct encounter with computer simulation dated back to his student days at the Department of Chemistry at Amsterdam University, in the early 70s, when he attended guest lectures by Berni Alder and later Les Woodcock. However, before that time he had already learned about computer simulations in statistical physics from reading a review on this topic by Berne and Harp in *Advances in Chemical Physics*. This review was studied by Daan and his fellow PhD students in a series of weekly evening sessions organized by his thesis advisor, Jan van der Elsken. His second, crucial meeting with simulation took place in 1975, when Daan was working, still at Amsterdam University, on his (experimental) PhD dissertation with Jan van der Elsken. In that year, Gianni Jacucci came to the department from CECAM to give a seminar on his recent work, together with Giovanni Ciccotti and Ian McDonald, on the "method of subtraction" in molecular dynamics. However, Jacucci not only gave a seminar on molecular dynamics; he also advertised that there was an opportunity for a young Dutch PhD student to spend a semester at CECAM in Orsay. In fact, this was not so much an offer as a request: the Dutch Science Foundation (ZWO) was one of the original contributors to CECAM, but at that time, they felt that in terms of research training they were not benefiting enough from their contribution. In the end, they got more than they bargained for.

Although other PhD students in the van der Elsken group had first choice (it was their "turn" to go abroad), none were able (or willing) to go to France on short notice (one week). When at last he was asked, Daan did not hesitate. The following week he left to spend five months in Orsay, where he would learn from Giovanni Ciccotti, Mike Klein, and Anees Rahman, and collaborate directly with Gianni Jacucci. On the basis of a copy of the MD code from the Verlet group, he was able to develop his own MD code. Jacucci made a tour of the US that summer and visited both Berni Alder and John Barker (who had carried out the first MC simulation of water with Bob Watts in 1969). Daan wanted to do event-driven MD simulations and hoped that Jacucci could get Berni's code. However, that was an idle hope, so Daan wrote his own. When visiting John Barker, Gianni Jacucci learned about John's path-integral based calculation of quantum virial coefficients. When he returned to Orsay, he asked Daan if he knew about path integrals. The answer was affirmative: Daan

---

[22]D. Frenkel, private communication addressed to G. Ciccotti, December 2018.

had just finished reading the work of Feynman and Hibbs during his holidays and was very enthusiastic about that approach. So Gianni and Daan then decided to try their hand at path-integral simulations of ideal bosons, fermions, and Boltzmann particles. This resulted in a report on path-integral simulations written by Daan for the 1975 CECAM Annual report. The work was never published: Daan had returned to Amsterdam where the computer power to continue was lacking. In addition, he realized that his real objective—computing spectroscopic properties by moving from imaginary time to real time—was beset with practical difficulties that could not be overcome in the few months he had left to finish his PhD research (in the end, it did also take others more than a few months).

Daan Frenkel leaving Amsterdam, bound to Orsay for his first visit to CECAM, June 1975

Once back in Amsterdam, Daan wrote a paper, based on his Orsay simulations (published in 1976), on the relaxation of quantized rotation of small molecules in dense, classical fluids. This topic was subsequently studied by other researchers, although initially it was often done using more simplistic approaches. In any event, ZWO's complaint to CECAM paid off, as Daan's stay at CECAM had borne its fruits: from that moment on Daan was working increasingly on computer simulation, a field that had gained recognition in the Netherlands (at least in chemistry—physics came later) through the work of Herman Berendsen at Groningen.

Meanwhile, Daan benefited from a grant to tour the US and Canada to visit the best research groups in his field. He was impressed by the various activities, but he was particularly enthusiastic about the open, collaborative atmosphere at the UCLA Physical Chemistry Department, where the intellectual climate was determined by people like Dan Kivelson, Bob Scott, Chuck Knobler, and John McTague, and a steady stream of inspiring guests, such as Paul Madden (then Cambridge, UK). Thus, when he got his PhD on the far infrared (now "terahertz") spectroscopy of dense fluids in 1977, he successfully applied for a postdoctoral position at UCLA. There, he started working, primarily with John McTague, a well-known experimental physical chemist, but also a computer simulator (who had collaborated with Rahman). But the boundaries between groups at UCLA were non-existent, so Daan's funding sometimes came from Dan Kivelson.

John McTague primarily saw Daan as a computer simulator. Daan used computer simulations as a tool to predict collision-induced Raman/Rayleigh scattering in simple molecular liquids. Inspired by McTague's growing experimental interest in the phase behaviour of quasi-two-dimensional systems of noble gases on graphite, and in particular by a lecture given by David Nelson in 1978, Daan started simulations of the melting transition in two-dimensional Lennard-Jones fluids, to search for evidence for the so-called "hexatic phase" that had just been predicted by Halperin and Nelson (earlier simulations by John Valleau predated the prediction of the hexatic phase, and hence he did not look for it). In collaboration with McTague, Daan produced the first results on this topic.[23] This paper and the ones that followed introduced a number of computational tools that were subsequently widely used: a practical prescription to compute bond-order correlation functions, and a scheme to visualize disclinations in two-dimensional systems.

After Los Angeles, Daan went back to Holland and obtained a job at Shell Research in Amsterdam. Daan enjoyed the work there. However, he came to the conclusion that he was working on projects that could be carried out much better by a skilled chemical engineer. Therefore, he looked for other options. He applied for, and obtained, a position as lecturer at the Physics Department of Utrecht University. In the five years Daan spent in Utrecht, he completed his transition from experimentalist to simulator. However, as a simulator, he retained his love for experiments. At this point, he became interested in computer simulations of liquid crystals, i.e. (mostly organic) substances which, under certain conditions, exhibit some properties

---

[23] D. Frenkel, J.P. McTague, *Evidence for an orientationally ordered two dimensional fluid phase from Molecular Dynamics calculations*, Physical Review Letters **42**, 1632, 1979.

of a fluid and some of a crystalline solid. At the beginning of the 80s, the experimental/theoretical study of real liquid crystals was expanding rapidly. Among the pioneers in liquid-crystal theory, we should mention the French scientist Pierre-Gilles de Gennes, who was awarded the Nobel Prize for Physics in 1991. «In Utrecht, I was involved in experiments on liquid crystals. I read de Gennes' book and I concluded that it would be a wonderful topic for simulations, which it was».[24]

However, Daan was not so much interested in molecular liquid crystals as in "lyotropic" (mainly colloidal) liquid crystals. His idea was to check under what conditions a simple system of thin, hard platelets (described approximately in the book by de Gennes) might undergo a transition to an orientationally ordered (nematic) phase. This ordering transition would only be driven by entropy. In fact, the idea of entropy-driven orientational ordering was not new: it had been proposed in the 40s by the Norwegian physical chemist (and Nobel laureate) Lars Onsager to explain the phase behaviour of rod-like colloids. And, of course, the freezing of hard spheres that was studied in 1957 by Alder and Wainwright was a prime example of entropy-driven ordering.

Together with, first an undergraduate student (Rob Eppenga) and later a (borrowed) PhD student (Bela Mulder), Daan started to explore the phase behaviour of hard non-spherical particles, mainly rods and plates. Apart from the observation (and characterization) of the isotropic-nematic transition in a system of thin hard disks (with Eppenga), the most important advance was the prediction of the shape-dependence of the phase behaviour of hard ellipsoids of revolution (with Mulder). This required a considerable extension of the free-energy calculation techniques that Daan had developed the year before with Tony Ladd. «In the early 80s there existed only a handful of publications aimed at computing complete phase diagrams, and these papers focused on simple atomic systems. Yet, atomic models do not begin to capture the phase behaviour of systems containing non-spherical particles. This is why the study of hard ellipsoids was so exciting.»

Just as Berni Alder had experienced before him, Daan found that many people had a hard time believing that increased order (e.g., the formation of a nematic phase) could be associated with an *increase* in entropy. For those who view entropy as a measure of disorder, these ordering transitions seemed to contradict the Second Law of Thermodynamics. However, entropy does not so much measure "disorder" as "freedom of motion", and entropic ordering occurs when the freedom of motion is greater in the ordered phase. This was confirmed in the mid-80s in a landmark experiment on colloidal hard spheres by Peter Pusey and Bill van Megen, while the experimental study of entropy-driven liquid crystal formation was considerably expanded, inspired in part by Daan's work on smectic and columnar phases.

---

[24]This and the following quote from D. Frenkel, private communication addressed to G. Ciccotti, December 2018.

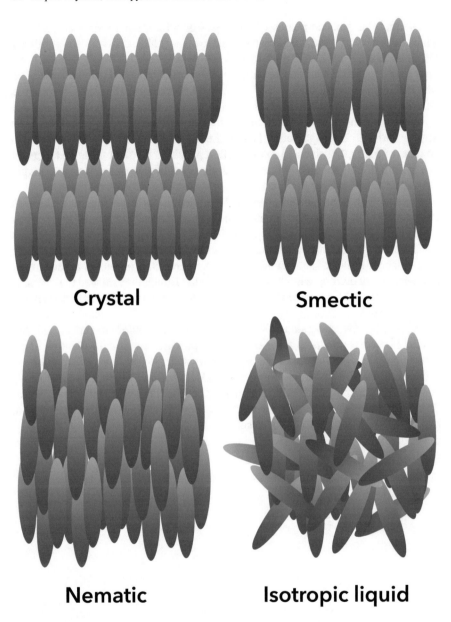

**Crystal**          **Smectic**

**Nematic**          **Isotropic liquid**

The different possible phases for a system of small ellipsoids, representing a liquid crystal, from the complete order of the crystalline phase (upper left figure) to the more disordered phase (isotropic liquid, below right); (M. Ferrario)

The fact that smectic phases (fluid phases of orientationally ordered particles, which form a layered structure) can form due to entropy alone had not been predicted. On the contrary, it had been argued that attraction was needed to form smectics. In his search for smectics, Daan was inspired by a new collaborator—Henk Lekkerkerker, who had just taken up a chair at the Van 't Hoff Laboratory of Physical Chemistry in Utrecht. «Daan had just started his simulation work on what is now known as entropy driven phase transitions», Lekkerkerker recalled. «Together with his students Bela Mulder and Rob Eppenga he worked on the isotropic-nematic phase transition in systems of hard ellipsoids and infinitely thin platelets. We then discussed whether in a system of rods hard-body interactions alone could give rise to the more highly ordered smectic liquid crystal phase».[25]

Together with Henk Lekkerkerker and his student Alain Stroobants, and later with PhD students Jan Veerman and Peter Bolhuis, Daan showed that, depending on their aspect ratio, hard rodlike particles could indeed form smectics (and hard platelets could form so-called columnar phases). This demonstration that entropic ordering may well be the rule, rather than the exception, forced many physicists to revise their identification of entropy with "disorder", since the simulations showed that in many systems order could increase in a closed system without heat transfer or other energy transformations.

Daan enjoyed working side by side with experimental colloid scientists, and considered himself fortunate that the experiments provided empirical validation of his simulation-based predictions. Daan would later state: «Now there is experimental evidence. There are real fluids—in the Van't Hoff Lab in Utrecht, among others—constituted by hard spheres, which crystallize in full accord with simulation. Nowadays, computer simulations may even predict how quickly crystallization goes on, and what kind of crystal form will result».[26]

---

[25]H.N.W. Lekkerkerker, *Daan Frenkel and the Spinoza Prize*, in: SIMU. Challenges in Molecular Simulations, Newsletter, Issue 2, November 2000, p. 1; https://www.researchgate.net/publication/267979729_SIMU_Challenges_in_Molecular_Simulations_Bridging_the_Length-_and_Timescales_gap_Volume_2.

[26]D. Frenkel, private communication addressed to G. Ciccotti, December 2018.

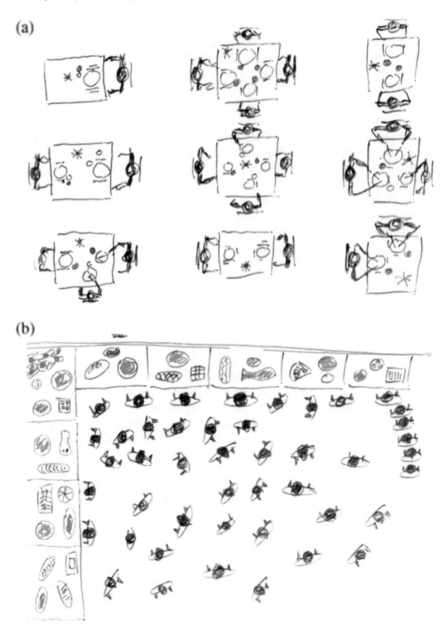

Daan Frenkel's "dinner party" sketch illustrating the different ordering criteria according to which a many-body system can be organized

The mid-80s were for Frenkel a period of great creativity and scientific productivity. He continued, with various collaborators, to produce results on the phase-transitions of various model systems, along the lines explored with Eppenga in the case of hard disks.[27] In 1984, together with Anthony (Tony) Ladd, he developed a new Monte Carlo simulation method in order to calculate the absolute value of the free energy of arbitrary solid phases.[28]

Tony Ladd was an English physical chemist who, after his PhD at Cambridge in 1978, moved to Davis University, California. Here, at the start of the 80s, he published a dozen papers on molecular dynamics, almost always with Bill Hoover, whom he joined at Livermore Laboratory in 1984. The crucial meeting between Frenkel and Ladd, which resulted in a published work, took place at a CECAM workshop in the summer of 1983. This workshop lasted almost a month, between June and July, and was devoted to the *Dynamics of Molecular Liquids and other Complex Systems under External Constraints*. As usually happened in these meetings, prolonged contact led people to discuss and work together along unforeseen directions.

Frenkel and Ladd resumed the idea of calculating the free energy of a state through thermodynamic integration, along a reversible path connecting the state of interest to a state of known free energy. For the solid phase, they chose as reference state an Einstein crystal, with the same structure as the solid under study (an Einstein crystal is a structure of atoms or molecules which do not interact among themselves and are constrained by elastic forces to their respective equilibrium positions.) The reversible path from the Einstein crystal to the solid under consideration may therefore be built by gradually "lighting up" the interactions of the model and "switching off"—still gradually—the harmonic oscillations; the fact that the crystal has the same structure as the real solid guarantees that, in this process, there are no sudden phase transitions, and that the path really is reversible. It can then be shown that the free energy of the initial solid phase differs from that of the reference Einstein crystal by a factor that is easy to calculate through simulation. The calculation can be done numerically through a Monte Carlo simulation on the ideal crystal, and deriving the free energy of the real crystal with a thermodynamic integration along a reversible path, characterized by the continuous variation of a parameter leading from the Hamiltonian of the ideal crystal to that of the real solid.

Once the broad lines of the method had been illustrated, Frenkel and Ladd applied it to calculate the free energy of a solid of hard spheres in the so-called fcc (face-centred cubic) configuration at the melting point. This was the same calculation carried out earlier on, using the single occupancy cell technique, by Hoover and Ree. Frenkel and Ladd's results agreed well with those obtained at Livermore. However, a much more versatile method had been developed, which could face the challenge

[27] D. Frenkel, J.F. Maguire, *Molecular Dynamics study of the dynamical properties of an assembly of infinitely thin hard rods,* Molecular Physics **49**, 503, 1983; D. Frenkel, B.M. Mulder, J.P. McTague, *Phase diagram of a system of hard ellipsoids*, Physical Review Letters **52**, 287, 1984; D. Frenkel, B.M. Mulder, J.P. McTague, *Phase diagram of hard ellipsoids of revolution*, Molecular Crystals and Liquid Crystals **123**, 119, 1985.

[28] D. Frenkel, A.J.C. Ladd, *New Monte Carlo method to compute the free energy of arbitrary solids. Application to the fcc and hcp phases of hard spheres*, Journal of Chemical Physics **81**, 3188, 1984.

of calculating the free energy for any solid phase. As the authors wrote in the last lines of the paper, the method was fast and accurate, and could also be extended to molecular solids.

Daan Frenkel and his research group, Amsterdam 1992. Almost all his young collaborators are today confirmed leaders in the molecular simulation community

## 6.4   Ab Initio Molecular Dynamics: Roberto Car and Michele Parrinello

«After Argonne, I came back to Trieste with a big pack of cards, and started to do some work with Tosatti, including molecular dynamics».[29] We left Michele Parrinello at Argonne Lab in 1980, just after the publication with Anees Rahman of the paper which extended to crystal solids the molecular dynamics method developed by Andersen.

Erio Tosatti was a theoretical physicist, who used to teach at the International School for Advanced Studies of Trieste (SISSA), where he worked on condensed matter physics. He was not strictly speaking a simulator, even though his research team produced a good number of papers on simulation. Nevertheless, Tosatti realized

---

[29]This quote and the following come from M. Parrinello, in D. Mac Kernan, cited in note 14.

the potentials of simulation techniques: «Tosatti was very open», Parrinello recalled. «He understood the power of Molecular Dynamics».

Parrinello had a position as associate professor at Trieste University, and often visited SISSA, located in Grignano, just outside the city. Here he met Roberto Car, who, after working for a while as a post-doc with Sokrates Pantelides at the IBM Thomas J. Watson Research Center, in the USA, was just back in Trieste as associate professor at SISSA.

Car was born in Trieste, but obtained his PhD at the Polytechnic Institute in Milan. He was two years younger than Parrinello, and had a different specialization: indeed, he was an expert in electronic structure. At SISSA, Car and Parrinello decided to work together. This came about in the month of December 1984.

Car had lots of ideas about how to improve understanding of the electronic structure of both atoms and molecules. On the other hand, Parrinello, too, was competent on both electrons and ions, beside statistical mechanics and molecular dynamics. Thus the two physicists decided to put together their complementary skills, thus combining electronic structure and molecular dynamics. «I think fortune and good luck are important: they put you in the right place at the right time», Parrinello said. Indeed, he was in the right place at the right time, and with the right person, in order to suggest and realize a quality leap in molecular dynamics. The problem they were trying to solve consisted in determining the potential responsible for the interactions in a many-body system, starting directly from the electronic configuration of its components.

As a general rule, the (more or less idealized, more or less complex) systems investigated in simulations are studied systematically by assuming the validity of what is known as the Born-Oppenheimer approximation. This provides a way to handle a cluster of atoms, interacting through the Coulomb forces between the various charged particles they are made up of (i.e., nuclei and electrons), considering only the nuclei, which are treated as classical particles "feeling" a mean field produced by the distribution of electronic charge, while the details of the latter are regarded as irrelevant. However, it is exactly the peculiar electronic configuration that produces the interaction potential. This difficulty had until then been bypassed by handling only systems of particles which interact in an extremely schematic way (such as hard spheres, or the liquid crystals studied by Frenkel), or by replacing the "real" potential with empirically derived phenomenological expressions. In most cases, these constitute a good enough approximation, giving satisfactory results in the description of the properties of the system. The most typical case, frequently used in simulation, is the Lennard-Jones potential.

However, there are situations, such as the systems held together by covalent bonds, in which the use of these interatomic empirical potentials is not adequate. Quantum theoretical chemistry has developed methods to derive, from the electronic distribution, the shape of the interaction potential, particularly with the so-called density functional theory (DFT). However, the corresponding calculations, although they provide an accurate description of the chemical bonds for a wide range of systems, are extremely costly from the computing point of view. As Car and Parrinello noted, this had until then prevented the application of DFT schemes to the study of very large molecules and/or systems of simple molecules, for direct calculation of the interatomic forces used in molecular dynamics simulations.

The qualitative leap made by Car and Parrinello consisted in managing to introduce in simulation the crucial information derived from the explicit handling of the electronic structure. As a consequence, a huge step forward was obtained in the ability of simulation to control the behaviour of a great variety of systems, reconstructing the interactions among the system components, starting from the fundamental level. At this point, the ab initio calculation which quantum chemists developed for the electronic structure of the single atom or molecule, or even that of a many-body system for fixed nuclear positions, also became a tool for ab initio simulation of the dynamics of many-body systems. The various more or less effective empirical potentials can be abandoned, and one can really start from the fundamental physical laws to reconstruct complexity.

Car was a DFT specialist, whereas Parrinello was by now an expert in molecular dynamics. In the end, the synergy between their complementary skills worked. Their attempt to create an effective computer program went on right through the cold winter in Trieste and the following summer in the USA, where Car went back to continue his collaboration with IBM. Those were months of hard work, first in Trieste, working also by night so as to get more machine time, and then in the USA, where Parrinello joined his colleague to continue to develop their ambitious idea.

In the end, all this hard work did produce the expected result. They wrote a paper, to be published in the *Physical Review Letters*, in which they proposed their new method. It was received well, even though the referees recommended a few adjustments. In a short time they made those changes and the paper which opened up a new field of studies to computer simulation was published.[30] Ab initio molecular dynamics was born. The abstract of the paper indicated clearly and without undue modesty the breakthrough achieved:

«We present a unified scheme that, by combining molecular dynamics and density-functional theory, profoundly extends the range of both concepts. Our approach extends molecular dynamics beyond the usual pair-potential approximation, thereby making possible the simulation of both covalently bonded and metallic systems. In addition it permits the application of density-functional theory to much larger systems than previously feasible».

Actually, extending this scheme to the simulation of metal systems remained for a long time more of an ambitious aim than a realistic goal. However, the progress made in the potentials of molecular dynamics was certainly very significant. More to the point, Car and Parrinello showed that the new method finally provided a way to calculate the properties of the electronic ground state of large and/or disordered systems at the state-of-the-art level of the calculations of electronic structure, in order to make ab initio simulations of molecular dynamics, assuming only the validity of classical mechanics to describe the ionic motions and the Born-Oppenheimer approximation

---

[30]R. Car, M. Parrinello, *Unified Approach for Molecular Dynamics and Density-Functional Theory*, Physical Review Letters **55**, 2471, 1985.

to separate nuclear and electronic coordinates. The method was successfully applied to the problem of calculating the static and dynamic properties of crystallized silicon.

Michele Parrinello (left) and Roberto Car (right) receiving from Berni Alder's hands the CECAM Prize in 2010 (CECAM archives)

Neither Car nor Parrinello were computer programmers. Therefore, their programme worked rather badly, as Parrinello himself admitted. In fact, in the original version, the rigorous ab initio calculation of the potential using DFT was only made in order to define the state of the system at the beginning of the simulation, and this was then completed without systematically re-calculating the potential at each step of the system evolution, as would be necessary to guarantee accurate results. Indeed, the original algorithm would be improved in the following years with the help of excellent programmers. However, the idea was definitely ground-breaking. The paper signed by Car and Parrinello and published in November 1985 met with immediate success and was soon used (and cited) by numerous researchers all over the world. After this success, the two scientists tried to apply the method to other realistic systems. «At this time», Parrinello recalled, «the first supercomputer came to Italy, it was a Cray and was put in Bologna, at Cineca. We had a very generous allocation of computer time and many able collaborators».

In that same year, Parrinello was appointed full professor at SISSA. In this school, created by Paolo Budinich a few years earlier (in 1978), the scientific atmosphere, as Parrinello recalled, was extraordinary. «There was a lot of freedom, a lot of free exchange of ideas. There was a group of very gifted people, Baroni, Tosatti, Selloni, and so on. There was no jealousy. We could talk freely. Everyone knew what the others were doing». All this freedom at SISSA was also due to the fact that researchers did not have to compete for a grant. They had enough time and resources, and this is what enabled them to think freely.

This ideal condition did not last long. In the space of three years, it disappeared. In 1989, Parrinello moved (broken-hearted, as he stresses) to IBM in Zurich, because SISSA could not assure him the technical support he needed in order to go on with his work (in other words, suitable staff and sufficiently powerful computers). Parrinello did not stay long at IBM, which was interested in more applicative research, so he moved once again and went to Germany, to the Max-Planck Institute for Solid State Research in Stuttgart.

Two years later, in 1991, Roberto Car, by now full professor, also left Trieste and went to work at Geneva University, and, at the same time, at the Romance Material Research Institute (IRRMA) of the École Polytechnique Fédérale de Lausanne, where he remained until 1999. He then moved to Princeton, USA.

Car and Parrinello were awarded the Dirac medal in 2009.

## 6.5 Gibbs' Ensemble and the "Blue Moon"

Athanassios Z. Panagiotopoulos was a PhD student working with Robert Reid at Boston MIT between 1982 and 1986. He was born and educated in Greece, but soon after his first degree at the Athens Polytechnic, he flew to the USA to obtain a PhD. Reid, he remembers, «was an exceptional classical thermodynamics specialist who loved the axiomatic conceptual approach and worshipped J. Willard Gibbs»[31]. In his first two years in he USA, he worked with Reid both as a theorist, on the empirical mixing rules of the equations of state, and as an experimentalist, working on the high-pressure behaviour of binary and ternary mixtures with carbon dioxide ($CO_2$) and water ($H_2O$). «But I realized after a couple of years that the future was in statistical mechanics and numerical simulations and taught myself Monte Carlo methods».

The first result of this change of scientific interests was a paper published in 1986 on the phase diagrams of mixtures of non-ideal fluids, simulated with the Monte Carlo method.[32] This work had several limitations, but nevertheless attracted the attention of Keith Gubbins, who, educated at London's Imperial College, had moved to the USA where he became director of the School of Chemical Engineering at Cornell University, in New York State, in 1983. In 1985, Gubbins took a sabbatical at Oxford, England. In Oxford he offered Panagiotopoulos a postdoc position. Once back at Cornell University, Gubbins changed the offer, offering the young Greek a junior faculty position, starting in Spring 1986. Panagiotopoulos accepted it, but arranged with Gubbins to spend a year in Oxford, England, as an independent researcher, nominally a guest of John Rowlinson, director of the Physical Chemistry Laboratory.

---

[31] This quote and the following come from A.Z. Panagiotopoulos, private communication addressed to G. Ciccotti, December 2018.

[32] A.Z. Panagiotopoulos, U.W. Suter, R.C. Reid, *Phase diagrams of non-ideal fluid mixtures from Monte-Carlo simulation*, Industrial & Engineering Chemistry Fundamentals **25**, 525, 1986.

«I arrived at Oxford—Panagiotopoulos recalled—in early September of 1986 with no clear idea on what I was going to do in the year to come. I had a vague notion that I wanted to optimize the determination of phase diagrams for fluid mixtures, broadly following up on the second half of my PhD thesis work».

Athanassios Panagiotopoulos

Quite often, we get brilliant ideas in unexpected places and moments. Some call them epiphanies. Or else, "Eureka moments". Well, Panagiotopoulos had this good epiphany while standing in the street. «The idea of setting up two boxes representing bulk phases and coupling them indirectly so as to satisfy the conditions of phase coexistence», he recalled, «came while I was waiting for a bus, on a rainy evening in late October 1986». This idea was inspired by the derivation of phase-equilibrium criteria proposed by Reid in his book *Thermodynamics and Its Applications*. Panagiotopoulos guessed that he had to follow this path. The bus arrived. He got on the bus, and once at home, went to his desk. «It took me a few hours to derive the acceptance conditions in the original paper, using purely classical arguments».

The method invented by Panagiotopoulos defined a new ensemble, in which he could obtain the phase-coexistence properties of fluid systems with more components using only one Monte Carlo simulation. The problem he hoped to solve by the new method had a direct application, inasmuch as the properties of phase equilibrium are the basis for a good number of separations used in industrial processes and affect the behaviour of a wide range of physical systems. At the same time, it was theoretically relevant, because given these properties, one can test the theories of the liquid state, and one can better assess intermolecular potentials for real fluids. The difficulty consisted in the complications emerging at the interface of the two phases in a fluid mixture, which make the simulations to find the appropriate equilibrium conditions and to control the chemical potential under these conditions extremely time consuming.

Panagiotopoulos' method bypassed the difficulty by eliminating the interface and inventing a sampling technique in a new ensemble. This allowed him to simulate the phase coexistence properties by following the evolution in space of the phases of a system separated in two distinct regions; the two regions generally have different compositions and densities, and are in thermodynamic equilibrium—both together and separately. The system is imagined immersed in an infinite medium, at constant temperature, with the total volume and number of particles fixed. During the system evolution, therefore, the kind of "moves" that can be made in the simulation, beside the conventional random displacement of a molecule within one of the regions, involve two further possibilities: a change of volume, equal and opposite in the two regions, which maintains equality of the respective pressures, and the random transfer of a molecule from one region to the other, to balance the chemical potentials of each component. For each of the three possibilities, the relative probabilities and conditions of acceptance are defined. The first type of move amounts to moving within the canonical ensemble (N,V,T) of the traditional Monte Carlo, the second operates in the (N,P,T) ensemble, and the third corresponds to a move in the phase space typical of the grand canonical ensemble, where fixed values are maintained for volume, temperature, and chemical potential.

Within a few days, Panagiotopoulos codified the method for pure Lennard-Jones liquids, and had the programme running on the Cray 1S supplied by the computer centre of London University. With this new generation of supercomputers, it was possible to sample a million configurations for a 500-molecule system in just five minutes of machine time. The programme worked. In fact, in the words of its inventor, it worked very well, and appeared «much more efficient than everything else available at the time». The young researcher immediately presented his result to Rowlinson and Gubbins, who realized its importance and encouraged him to publish it as soon as possible in *Molecular Physics*. Easier done than said in this case! Panagiotopoulos submitted his paper on December 22 and, after a quick revision by Dominic Tildesley, from Southampton, the paper was published in a very short time: a month.[33]

The name, Gibbs ensemble, as it was announced in the original paper, had been chosen because «the equations posing the necessary and sufficient conditions for phase equilibrium between the two regions I and II had been originally derived by J.W. Gibbs, more than 100 years ago». It was somehow an inappropriate name, because it had been used earlier on, in order to indicate another type of ensemble, called the *generalized ensemble*. «But I did not want to leave open the possibility that the method would come to be known as "Panagiotopoulos Monte Carlo" and I wanted to (indirectly) honor my advisor, Bob Reid, for his influence».

Athanassios Panagiotopoulos, on a suggestion of Gubbins, later wrote a second paper on the method he had developed, applying it to the phase transitions of pure fluids in pores.[34] Gubbins had previously worked on this problem. Meanwhile, at the

---

[33] A.Z. Panagiotopoulos, *Direct determination of phase coexistence properties of fluids by Monte Carlo simulation in a new ensemble*, Molecular Physics **61**, 813, 1987.

[34] A.Z. Panagiotopoulos, *Adsorption and capillary condensation of fluids in cylindrical pores by Monte Carlo simulation in the Gibbs ensemble*, Molecular Physics **62**, 701, 1987.

beginning of 1987, the referee of his first paper, Dominic Tildesley, came out into the open and—together with his PhD student Mike Stapleton and Nick Quirke, who worked at BP Research—, contacted Panagiotopoulos and asked him whether one could use his algorithm when working on mixtures. The answer was positive. This led to his third paper, which extended his method to "osmotic" equilibria.[35]

By then, Panagiotopoulos had gone back to the USA and become junior faculty at Cornell. Stapleton, once he had completed his PhD, accepted to join him in January 1988. He was his first post-doc student.

The problems of physical chemistry arising in the study of the properties at the interface of different systems were one aspect of the growing complexity of the situations which simulation was starting to face and solve in that period. Another aspect of the same complexity was tied up with the possibility of simulating on a computer the behaviour of phases of matter in which chemical reactions are taking place.

The theoretical interest in chemical reactions was long-standing, and was connected to some of the great names of statistical mechanics, such as Lars Onsager. Even in the early 60s, a significant theoretical contribution to the problem of calculating relevant parameters for this kind of reactions had been given by Tsunenobu Yamamoto (the same who, twenty years later, would supervise the PhD work of Shuichi Nosé)[36]; further suggestions towards a formulation of the problem, which might have been effective from the computing point of view, had come from James Keck, [37] and were resumed, about ten years later, by James Anderson.[38] The emerging idea was that it would be convenient to study the problem by considering the system close to the transition (i.e., close to the top of the barrier separating the two stable states), and from that point calculating the transmission coefficient, i.e., the probability that the reaction may effectively take place. Anderson stressed the fact that this procedure might be particularly effective on computational grounds, inasmuch as, by starting the trajectories near the transition barrier, one would not waste time in calculating paths that are irrelevant for the reaction kinetics. The point was that the barriers involved in the reaction processes create bottlenecks, so that the transitions from one stable state to the other are "rare events", which take place with a frequency several orders of magnitude lower than those typical of the molecular motions that would be followed in direct molecular dynamics simulations.

[35] A.Z. Panagiotopoulos, N. Quirke, M. Stapleton, D. J. Tildesley, *Phase equilibria by simulation in the Gibbs ensemble: alternative derivation, generalization and application to mixture and membrane equilibria*, Molecular Physics **63**, 527, 1988.

[36] T. Yamamoto, *Quantum Statistical Mechanical Theory of the Rate of Exchange Chemical Reactions in the Gas Phase*, Journal of Chemical Physics **33**, 281, 1960.

[37] J.C. Keck, *Variational Theory of Chemical Reaction Rates Applied to Three-Body Recombinations*, Journal of Chemical Physics **32**, 1035, 1960.

[38] J.B. Anderson, *Statistical theories of chemical reactions. Distributions in the transition region*, Journal of Chemical Physics **58**, 4684, 1973.

As we saw early on, the question of the different time scales within complex events of this kind, and the need to develop methods able to study rare events, somehow "forcing" the system to cross areas of low probability, had already been the focus of the attention and of the first pioneering attempts by Charles Bennett. On a more strictly theoretical level, the same problem had been faced a couple of years later by David Chandler, a physical chemist educated at Harvard (where, for some time, he had collaborated with Andersen on problems in the statistical mechanics of fluids), who provided the first formal justification of Bennett's method, and, in general, a solid foundation, on the basis of the fundamental principles of statistical mechanics, for attempts to define rules for effective sampling, as well as dynamic criteria for a correct calculation of the reaction speed in systems activated in a liquid phase.[39]

Therefore, in the early 80s, one could see on the table a theoretical formulation of the problem, on the basis of the fundamental principles of statistical mechanics, as well as the difficulties and the best strategies to adopt to translate it effectively into a simulation.

In the summer of 1985, CECAM organized at Orsay a first workshop devoted to a stochastic approach to chemical reactions, in an attempt to clarify the ideas which were circulating on that issue. The idea of this workshop originated from the meeting between Ciccotti, who had been working for some time on the problem of constraints, and Pierre Turq, a physical chemist from the Université Pierre et Marie Curie in Paris, who found at CECAM an environment well tuned to his personal interest in simulation, which was at the time still considered a "barely respectable" research area in the academic community of theoretical chemistry in France. Among the workshop participants were Ray Kapral and James (Casey) Hynes, from Toronto and Boulder, respectively. The two scientists had come to know each other well during the years spent together at Princeton, working on their PhDs in physical chemistry. Kapral got his PhD in 1967, whereas Hynes obtained it two years afterwards. Later on, in 1969, Kapral got a position as assistant professor at the University of Toronto, while Hynes found a permanent position at the University of Colorado. At Orsay, they started discussing with Ciccotti, one of the workshop's organizers, the theoretical questions related to the determination of the velocity correlation functions in chemical reactions. They had the idea that they might use the theoretical apparatus of statistical mechanics to calculate the speed of chemical reactions explicitly by simulating the dynamics of a reagent system on a computer.

---

[39] D. Chandler, *Statistical mechanics of isomerization dynamics in liquids and the transition state approximation*, Journal of Chemical Physics **68**, 2959, 1978.

Pierre Turq (1943–2015)

Around the mid-80s, this topic was on the agenda: indeed, it was resumed in three following CECAM summer workshops, from 1986 to 1988, all of them dedicated to the computer simulation and the theory of chemical reactions in solution. Gradually, a working group was formed among the various participants at these meetings, constituted by Ciccotti, Hynes, Kapral, and Mauro Ferrario. They began to study the formation of ion pairs in polar solvents as a simple case to approach the general problem of reaction in a condensed phase. The theoretical handling of the problem in statistical mechanics involves the evaluation of the potential of the mean force, and of the dynamic effects linked to overcoming the potential barrier that must be "crossed" for the reaction to take place. In a couple of papers devoted to the evaluation of the potential of the mean force, the idea arose to introduce constraints enabling the molecular dynamics simulation to force the system to pass through the relevant values of the reaction coordinates.[40]

The final formulation of the method took place between 1987 and 1988. Ciccotti was taking his leave from Rome and spending his sabbatical between Toronto and Boulder. In the last stage of the work, Emily Carter joined the Ciccotti-Hynes-Kapral trio. Emily had just obtained her PhD in physical chemistry at Caltech, and had won a postdoc grant at Boulder, where she started working with Hynes on the dynamics of electron transfer in solutions.

As we saw earlier, reaction dynamics typically involves crossing a free energy barrier, and this leads to rather long reaction time scales, so that reactive events are quite rare. The introduction of suitable constraints may force the system to stay near the bottleneck, but this forced limitation, though it made it possible to simulate the phenomenon without the excessive computing costs due to time spent in visiting irrelevant areas, may lead to significant alterations in the distribution of configurations, and, ultimately, to misleading, or non-physical results. The method developed by the four authors avoided this difficulty thanks to a technique for "re-weighting" the altered configuration distribution in an appropriate ensemble, namely *constrained reaction coordinate dynamics*. The final paper, in which the new method was fully explained, was published in the month of April 1989.[41] «A computationally efficient molecular dynamics method for estimating the rates of rare events that occur by activated processes is described [...] This new method provides an independent alternative to the commonly applied umbrella sampling techniques».

---

[40]G. Ciccotti, M. Ferrario, J. T. Hynes, R. Kapral, *Molecular Dynamics Simulation of Ion Association Reactions in a Polar Solvent*, Journal de Chimie Physique **85**, 925, 1988; G. Ciccotti, M. Ferrario, J. T. Hynes, R. Kapral, *Constrained Molecular Dynamics and the Mean Potential for an Ion Pair in a Polar Solvent*, Journal of Chemical Physics **129**, 241, 1973.

[41]E.A. Carter, G. Ciccotti, J.T. Hynes, R. Kapral, *Constrained Reaction Coordinate Dynamics for the Simulation of Rare Events*, Chemical Physics Letters **156**, 472, 1989.

Giovanni Ciccotti with a different "blue moon" (photo by K.E. Gubbins)

The title of the paper was anodyne, and meant little to non-specialists. Even specialists found it difficult to pronounce the definition "constrained reaction coordinate dynamics", even though they knew what it meant. In a footnote, the authors suggested an alternative; since the CRCD ensemble referred to rare events, which took place only *once in a blue moon* (a typical English expression to describe rare events), an appropriate definition might in fact be a *blue moon ensemble*. And *blue moon* was indeed the name that survived.

# Chapter 7
# Quantum Systems and Critical Phenomena

We have seen how, in the first thirty years following its birth and with the first encouraging results obtained through the study of simple model systems, molecular simulation had grown in all possible directions, imposing the effectiveness of its own methods for a complete realization of the programme of statistical mechanics: derivation and explicit calculation, starting from the basic building blocks (i.e., atoms) and from the fundamental laws of physics, of the properties of the aggregate states of matter in its various forms, and of the transitions between them.

The developments we have so far described, however, are all limited to the area of classical statistical mechanics; indeed, while the microscopic building blocks of the systems under study are—in principle—described by quantum mechanics, for the study of their collective behaviour we may use approximations to reduce the problem to a strictly Newtonian dynamics. Indeed, the remarkable mass difference between electrons and nuclei means one can decouple the system into a set of massive nuclei, immersed in a field resulting from the ground state of the electronic distribution, depending parametrically on the nuclear configuration. At a normal temperature, if the nuclear de Broglie wavelength is small in comparison with the typical internuclear distance, the nuclear quantum properties do not have significant effects on the behaviour of the system, so one can simply study a group of classical particles interacting through a potential determined by the electronic distribution, provided it is adequately modelled. This is still true when, as it happens with the method developed by Car and Parrinello, the effect of the electrons upon the interaction between nuclei, of entirely quantum nature, is taken into account *ab initio*, by explicitly calculating the electronic ground state to assess the interacting potential for each nuclear configuration.

We are left with the problem of the feasibility of computer simulation programmes which explicitly take into account the quantum nature of the system, therefore facing the specific challenges of quantum statistics. Such challenges cannot be ignored if we wish to switch to studying an electron gas, or the superfluid transition of low-temperature helium. The first few steps in the progress of quantum simulation were sketched fairly early on, but it was only from the early 80s onwards that the field of quantum statistics was embedded within simulation techniques. The breakthrough

© Springer Nature Switzerland AG 2020

G. Battimelli et al., *Computer Meets Theoretical Physics*, The Frontiers Collection,
https://doi.org/10.1007/978-3-030-39399-1_7

for simulation in the field of quantum statistics was largely due to the work of the US physicist David Ceperley. However, we should add that, as always happens in the development of a research field, linking a remarkable result to the work of one scientist alone might prove overly reductive, and unfair to others who may have contributed to it; in this particular case, Malvin Kalos played an important role as a forerunner.

In the same way, the success obtained in making the so-called "critical phenomena" accessible to simulation should largely be credited to the work of the physicist Kurt Binder, of Austrian origins. This category of phenomena, all within the confines of classical statistical mechanics, had long been rather deliberately avoided by the founders of simulation, due to both conceptual and computational difficulties. Critical phenomena occur close to particular values of the thermodynamic parameters ("critical points"), featuring peculiar properties which were barely understood by theoretical physicists at the end of the 60s (let alone amenable to rigorous analytical results).

The two following figures allow us to visualize the situation. The first shows a typical phase diagram in the pressure-temperature plane, with the domains of (co)existence of the solid, liquid, and gaseous phases. The $P_C$ point, at which the liquid and gaseous phases merge, is the critical point. In its vicinity, phenomena take place that still escaped theoretical understanding and that would be the focus of Binder's attention in simulation. The second figure is a sort of "zoom" on the first, at a low temperature, and shows the typical quantum effects associated with the transition to the superfluid phase of liquid helium when the temperature is lowered so much that it is no longer possible to assimilate helium nuclei to classical material points, whence their quantum nature manifests itself through the phenomenon of Bose-Einstein condensation. This is the kind of phenomenon that Ceperley's work on quantum simulation would be able to control and calculate.

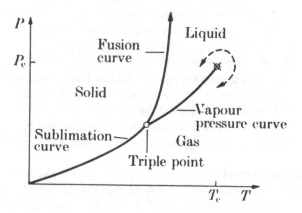

*Source* H.E. Stanley, introduction to phase transitions and critical phenomena, Oxford University Press 1971

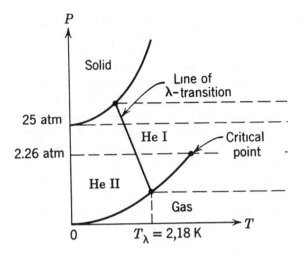

*Source* K. Huang, statistical mechanics, Wiley 1963

In the following sections, we shall trace the parallel paths followed by Ceperley and Binder to handle quantum problems and critical phenomena computationally. These two scientific paths led to crucial results which helped remove the last traces of resistance towards simulation as an essential tool for the development of theoretical physics among those physicists concerned with the study of matter in all its forms.

## 7.1  Mal Kalos, David Ceperley, and Quantum Simulation

In 1972, in the review article already mentioned, Berni Alder considered the shift to the simulation of quantum systems as one of the open frontiers of simulation, underlying the still embryonic stage of this research field. In the quantum case, the first problem to be solved was the calculation of the ground state, eigenstate of the time-independent Schrödinger equation, which contains all the information about the state of the system. However, the related conceptual difficulties and, all the more decisively, the limited performances of the first computers, had confined early attempts to mere declarations of intent. Around the mid-70s, the higher computing power available enabled scientists to make more serious attempts to face the problem.

The technique used in these first attempts at quantum Monte Carlo is known as Variational Monte Carlo (VMC). Its broad lines, fixed in a paper by William McMillan of 1965[1], are the following. One starts from the energy of the system, as obtained from a trial wave function, for which an explicit form is assumed in terms of the particle positions and a few variational parameters. It is well known that the ground state of the system corresponds to the minimum of the energy. Leaving those parameters fixed, one calculates the energy and its derivatives with respect to the variational

---

[1] W.L. McMillan, *Ground State of Liquid He$^4$*, Physical Review A **138**, 442, 1965.

parameters, using a Monte Carlo calculation in which the probability of configurations (particle positions) is just the square modulus of the wave function. One changes the value of the variational parameters in the direction of the energy derivatives with respect to the parameters and repeats the Monte Carlo calculation. One continues this iterative process until the minimum for the energy is found, obtained when the derivatives are null. The final trial function is the best approximation to the wave function of the ground state within the chosen functional form.

This method is obviously approximate, inasmuch as it is based upon a choice of the shape of the trial function. To obtain the ground state of a quantum system there are at least two major methods. The first appeared very early, even before the appearance of Variational Monte Carlo; this is Green Function Monte Carlo, based on the iterative solution of the time-independent Schrödinger equation written in integral form.[2] The second, (numerically) exact method, is called Projection Monte Carlo. Here the ground state is obtained by evolving the system in imaginary time, which amounts to evolving a stochastic process in the configuration space, thus getting a sample of the ground state as the asymptotic evolution of the sample extracted from the initial state. Apart from possible computational difficulties, this allows one to draw relevant physical properties with average operations using the sample of the state which has thus been created.

Malvin Kalos stood out among the protagonists of the pioneering stage of quantum simulation. After getting his PhD at the University of Illinois in 1952, he worked for several years, first as a post-doc student with Hans Bethe at Cornell University, and later on as an expert in simulation for a private company, the United Nuclear Corporation. In these first years of scientific activity, as he studied the diffusion of neutrons and related nuclear properties, Kalos realized that he might handle Schrödinger's equation for an N-body system, interacting with a pair potential, essentially like a random walk in a 3 N-dimensional space, and solve it effectively with the Monte Carlo method.

---

[2]M.H. Kalos, *Monte Carlo Calculations of the Ground State of Three- and Four-Body Nuclei*, Physical Review **128**, 1791, 1962.

Malvin Kalos

In 1964, Kalos left the private sector and accepted an academic position at the Computing Department of the Courant Institute of Mathematical Sciences of New York University. Here he was directly concerned with the quantum Monte Carlo method, and was stimulated in particular by Loup Verlet, who had been invited to New York by Joel Lebowitz. In 1970, Verlet invited him to Orsay to give a series of lectures on quantum Monte Carlo, and a close collaboration sprang up between the two scientists. This resulted in an important paper, published in 1974, where simulation with the Monte Carlo method was extended to a quantum system, calculating the properties of the ground state of a hard-sphere system (obeying the Bose-Einstein statistics) as a model for liquid helium.[3] They used the Green Function Monte Carlo (GFMC) method, a generalization to quantum mechanics of the classical Monte

---

[3]M.H. Kalos, D. Levesque, L. Verlet, *Helium at zero temperature with hard-sphere and other forces*, Physical Review A **9**, 2178, 1974.

Carlo technique, based upon the fact that, to a first approximation, the wave function of the ground state for a boson system may be taken as real and positive.

The interaction between Kalos and the Orsay group produced a secondary effect, which, as we will see, would have important consequences. Jean-Pierre Hansen got a post-doc position at Cornell University through Geoffrey Chester, who, through his mediation, began a collaboration with Kalos in New York. In order to consolidate their collaboration, Chester suggested that his PhD student, David Ceperley, start working with Kalos.

Let us come back to the second set of "exact" methods, those called Projection Monte Carlo. These methods use a projection depending on the evolution in imaginary time, i.e., the evolution operator is applied to the initial state. This operation, applied to the trial function, transforms it (for a long enough evolution) into the exact ground state to be determined. The asymptotic evolution of this pseudo-dynamic problem does indeed solve Schrödinger's equation in an imaginary time. This amounts to handling a diffusion equation in the presence of interactions, an equation which may be solved with stochastic trajectories (random walks). By sampling the density corresponding to the initial trial state (which is just the best trial function of VMC) an ensemble of representative configurations is created. Starting from the elements of this ensemble, suitable random walks are created, which sample the evolution of the trial function at the imaginary time t. After a sufficiently long time (which depends on the energy difference between the ground state and the first excited state), the evolved members of the ensemble will be distributed according to the probability density of the ground state. With this sample, one can calculate the average of observables upon this state.

The various options available to achieve this scheme effectively in a simulation (how to perform random displacements, and above all how to assign probabilities for the successive steps of the procedure) give rise to variations on the theme, which are known as Diffusion or Reptation Monte Carlo. These are different ways to optimize the same fundamental scheme, taking into account the fact that competing parameters are often at stake. For example, if we reduce the interval of discrete time between the successive steps of the diffusion process, this enables us to cut down errors in calculations, but noticeably increases the number of necessary operations, thence the machine time required for the final result. Otherwise, more elaborate techniques for suitably evolving the trial function, although time-consuming, can nevertheless lead to a quicker convergence of the process, thus saving, after all, the overall computational cost of the operation.

As we have anticipated, David Ceperley was the one who took the lead in the development of these methods. Ceperley was born in 1949 in Charleston, West Virginia; after obtaining a Bachelor Degree in physics and maths at the University of Michigan, he started studying for his PhD in physics at Cornell in 1971, under the guidance of Chester, in a lively environment, where he met leading scientists, such as Ken Wilson, Hans Bethe, David Mermin, and Michael Fisher. In his third year of studies, he was put directly in contact with Malvin Kalos through Chester. Ceperley's PhD dissertation would be completed, between Cornell and New York University (Courant Institute), under the joint supervision of Chester and Kalos.

David Ceperley (courtesy Department of Physics, University of Illinois at
Urbana-Champaign)

The task Kalos gave Ceperley for his dissertation consisted in the generalization
of the Monte Carlo method to fermion systems, i.e., systems composed of spin
half particles—obeying Fermi-Dirac statistics and subjected to Pauli's principle of
exclusion. The anti-symmetry property of the wave function of such systems involved
remarkable difficulties for simulation. On the one hand, there were basic calculation
difficulties: the wave function of the ground state of a fermion system may be either
positive or negative (even complex), and this made it impossible to handle it directly
as a probability, unlike the boson case, in which there was no such problem. There
was the need to develop techniques, and build algorithms, which could provide a way
to suitably handle, and effectively sample, quantities which might take up negative
values (the well-known "sign problem" which still puzzles simulators working on
Monte Carlo for fermions even today). On the other hand, serious difficulties turned
up, related to calculation time: the fermionic wave function required (at least) the
calculation of a determinant, and this increased the number of necessary operations

to a point that was considered prohibitive by many. Ceperley started facing these problems, thus becoming acquainted with simulation techniques and developing algorithms which allowed him to obtain a few approximate results, and these were summed up in his PhD dissertation, submitted in the summer 1976.

Soon after, he flew to Orsay with a 1-year post-doc contract, attracted by the perspective of working with Loup Verlet. However, as we mentioned earlier on, Verlet had retired from physics, and Ceperley worked with the other members of the group. Because of the lack of interactions between this group and CECAM, Ceperley missed the chance to get in touch with the group of simulators which, in the meantime, was growing on the hill at Orsay. Ceperley was looking for further applications of his dissertation work, and, encouraged by Hansen, focussed upon a simple system based on a fermion model, i.e., a gas of electrons interacting with a Coulomb potential. The results of these Monte Carlo simulations, realized at Orsay to determine the equation of state of a uniform electron plasma in two and three dimensions, were published the following year.[4] During his stay in Paris, Ceperley collaborated with Kalos, who visited Paris in the spring of 1977, in drafting a review article on the application of the Monte Carlo method to quantum systems. At the time, this represented a summary of everything then known on quantum simulation.[5] It was published in a collective volume, edited by Kurt Binder, which reviewed the use of the Monte Carlo method in statistical physics. In the same period, Ceperley started getting interested in developing effective simulation methods for polymers, upon a suggestion of Lebowitz, who was also in Paris on his sabbatical.

At the end of his stay in France, Ceperley met Berni Alder. They got acquainted during a summer school dedicated to the microscopic and dynamic structure of liquids, near Aleria, in Corsica in September 1977. The two physicists talked to each other, and Berni was probably struck by young David. The point is that Alder showed interest in Ceperley's studies, and appreciated his skills. He thus started trying to find a way to bring Ceperley with him to California.

After his short stay in Paris, Ceperley went back to the USA. Over the next year he oscillated between two places: Rutgers University, with Joel Lebowitz, for a computer simulation of polymer behaviour, and New York, where he continued to collaborate with Kalos on the fermion problem. In the summer of 1978, he finally arrived in Berkeley, where, a year earlier, the National Resource for Computational Chemistry (NRCC), a centre of theoretical chemistry, had been established. Ceperley got a semi-permanent position at NRCC, as an expert in statistical mechanics. Ceperley and Alder started working together. At this point, the scientific relationship between them became concrete.

In fact, this partnership developed in precarious, yet lucky conditions. Indeed, in order to develop his research, Ceperley needed a computing power which only Livermore computers could guarantee. To get access to these computers, David

---

[4]D.M. Ceperley, *Ground state of the fermion one-component plasma: A Monte Carlo study in two and three dimensions*, Physical Review B **18**, 3126, 1978.

[5]D.M. Ceperley, M.H. Kalos, *Quantum many-body problems*, in *Monte Carlo Methods in Statistical Physics*, ed. by K. Binder, Springer-Verlag 1979.

needed a security clearance, which the security authority only released after a year. However, he did not lose heart, and found the right trick: he started working by phone. In practice, Ceperley would visit Livermore several times a month to plan activities, but in-between he gave instructions to Mary Ann Mansigh, Alder's collaborator, who, in turn, got the Livermore computers "running" for the assigned time. The following day, she would call him back to tell him which results had been obtained. This procedure was rather cumbersome, but productive all the same.

Alder had long been interested in the problem of quantum simulation, and knew David's teacher, Malvin Kalos, and the topic he was studying. Indeed, he had met him at least fifteen years earlier. «I knew that Kalos was working on the quantum Monte Carlo problem. I sat next to him every day, it must have been the 60s, I really tried to understand his algorithm, but it was so complicated, and he could not explain it to me».[6] Perhaps he was unable to explain it, or perhaps he just did not want to. Since Alder added: «Quantum Monte Carlo was made clear to me by Ceperley», several years later. However, he underlined: «Well, to be fair to Kalos, quantum Monte Carlo had progressed a lot by then, and was much clearer. Anyway, Ceperley had a clear understanding of quantum Monte Carlo, and we started talking» . As mentioned earlier on, the discussion had started in Corsica. Now in California, however, they did not only talk, they actually started working, since they could use a very powerful computer, Cray 1.

As Alder recalled later on, Ceperley «had worked on the electron gas, by an approximate method, variational Monte Carlo. I said, David, no one is going to believe your results, unless we do it exactly, without approximation. That's how we got started on the electron gas, and we did it exactly eventually. It took thousands of hours of the fastest (Cray 1) computer. We just bullied our way through, both intellectually, and numerically, and we got a classic paper [...] It was a very nice job, and David was certainly the intellectual driving force behind it».

The paper Alder referred to is *Ground State of the Electron Gas by a Stochastic Method*, published in the month of August 1980.[7] «The most important paper in condensed matter physics at the time», Berni Alder defined it, without false modesty. «My best work», according to Ceperley.[8] As was summed up in the first few lines of the paper, for the first time the properties of the ground state of a quantum system were calculated through an exact stochastic simulation of Schrödinger's equation for charged bosons or fermions. In particular, they underlined that the Monte Carlo method they used, «if run long enough on a computer, can give as precise a solution for the ground state of a given fermion system as desired» .

---

[6]This quote by Alder, and the following, can be found in D. Mac Kernan et al., *Interview with Berni Alder*, SIMU Newsletter n.4, 2002, https://doi.org/10.13140/2.1.2562.7843, https://www.researchgate.net/publication/267979976_SIMU_Challenges_in_Molecular_Simulations_Bridging_the_Length_and_Timescales_gap_Volume_4.

[7]D.M. Ceperley, B.J. Alder, *Ground state of electron gas by a stochastic method*, Physical Review Letters **45**, 776, 1980.

[8]This quote by Ceperley, and the following, unless otherwise stated, come from D. Ceperley, *A Quantum History,* personal communication to the authors, June 18, 2018.

Berni Alder and David Ceperley in a relaxed moment during a conference
on many-body physics in Mexico, January 1981 (D. Ceperley)

The 1980 paper merged the results obtained by Ceperley in the approximate
simulations of the previous year, as well as the huge amount of work he had done
in the year he had spent at Berkeley. In this period, thanks to direct interaction with
Mary Ann Mansigh, Ceperley had become an expert in creating fine-tuned codes
which could run on Cray-1 to minimize errors and calculation time, and he had
developed and implemented in effective algorithms the ideas for a proper treatment
of the peculiarities of the calculation of the ground state of a fermion system, i.e.,
the sign problem, arising from the anti-symmetry of the wave functions.

This was a version of Diffusion Monte Carlo, in which one starts with a trial
wave function (either positive or negative) which one would like to sample and
evolve (i.e., asymptotically project upon the ground state). The procedure consisted
in Monte Carlo sampling an ensemble of configurations from their probability, given
by the initial state, and letting them evolve. If the initial wave function, calculated

in the evolved configuration changed its sign, that state was discarded. Otherwise, it could be held (this hypothesis was called "fixed nodes"). The sample obtained in this way represented a faithful sample of the ground state, and could therefore be used to calculate the average values representing the system properties. In fact, one was excluding configurations corresponding to zero value of the evolved quantum state which corresponded to an approximately zero probability. This source of error could be evaluated and corrected with a further procedure.

These methods, known as "fixed-node" and "released-node", as well as their related codes, were developed by Ceperley and discussed with Alder between September 1978 and May 1979. During the spring of 1979, the programme started running on a computer and gave its first results. The paper which summed them up was published the following year; in the final lines, beside the credits, usual in Alder's papers, given to Mary Ann Mansigh for her assistance in the computation stage, the authors explicitly acknowledged their debt to Malvin Kalos, «for numerous useful discussions», but mostly «for inspiring the present work».

The paper by Ceperley and Alder showed the huge potentials of the Quantum Monte Carlo method and gained immediate resonance, above all thanks to the impact it had upon the community of physicists, who were working on the density functional theory (DFT) to compute the electronic configuration. From then on, it remained a milestone: in 2003, in a list of the most significant papers published in *Physical Review Letters*, written on the occasion of the fiftieth anniversary of this journal, this paper was in third place as regards the number of citations.

Provided one gets a great deal of machine time, Alder and Ceperley stated, one can achieve a precise solution of the problem of the ground state, not only for an electron system, but also for any fermion system. It is therefore natural that the next step was the application of this method to a more realistic problem than the electron gas model. «After I finished the electron gas calculations», Ceperley recalls, «with Berni's urging, I began to work on many-body hydrogen in 1980. An electron gas is not directly realized in any material, it's an idealized model, while hydrogen is a real material. With the hydrogen calculation we wanted to address experimental predictions, not just compare with theory. Our hydrogen calculation was the first many-electron calculation of a material to lead to important predictions».

Interest in the properties of high-pressure hydrogen was generally quite strong at Livermore, because of its relevance for the calculations of thermonuclear explosives and for the first studies on inertial-confinement fusion. Alder had worked on this kind of problem since his arrival at the laboratory.

The elements needed to face this challenge were ready. There were a few complications to be solved, related to the fact that the system under study was an aggregate of two fermionic components, one of which (i.e., protons) was about two thousand times more massive than the other. This situation featured a double scale of the pseudo-dynamics used in the method followed for the calculation. Protons "move" much more slowly than electrons, and since the elementary integration step is given by the electrons' "motion", one has to follow the system for thousands of steps before protons move significantly. As a consequence, there was a really heavy demand on computing time: «You may need to have millions of time steps before the proton's distribution will converge». On the other hand, Alder and Ceperley were interested in studying the high-pressure phase properties of hydrogen, in particular the transition between the metallic and molecular phase, and it is the proton distribution that affects the physical phase of the system. It was therefore necessary to build trial functions including not only electron-electron and electron-proton terms, but also proton-proton correlations. «These calculations were at the cutting edge of what could be done in the early 1980's». The first preliminary results were published in 1981, but simulations on hydrogen went on for a long time after that, and a complete study was only published in 1987.[9]

## 7.2 Applying Feynman's Path Integrals to Simulation: Superfluid Helium

In 1981, NRCC was closed. This was a sudden, probably mistaken decision, since, according to Ceperley, the centre might still have helped US chemistry. However, this was a good opportunity for David, because he was offered a permanent position at Livermore Laboratory. Obviously, he accepted. Thus he could go on working with Berni Alder.

At Livermore, Ceperley established a new collaboration with Roy Pollock, a young scientist with a profile extraordinarily similar to David's. Indeed, Pollock got a degree at Cornell just like Ceperley, then crossed the Atlantic Ocean in order to work as a post-doc in France, at Orsay. Later on, he went back to the USA and obtained a position at Livermore. In this period, he worked with Alder's group, and this allowed Pollock and Ceperley to collaborate. They focussed upon the study of the properties of superfluid $^4$He.

To explain what it's all about, we should remind that helium is a noble light gas. After hydrogen, it is the most common element in the universe. However, it is also the only element which does not solidify at ambient pressure, and remains liquid even close to absolute zero. $^4$He is the helium isotope with a nucleus made up of two protons and two neutrons. The feature which really makes it attractive to physicists is that, below 2.1 K, it becomes superfluid: in other words, it manages to isolate itself from surrounding matter, and flow without viscosity. Its superfluidity is the result of the so-called Bose-Einstein condensation (BEC): at low temperature, the system behaves like one coherent state.

However, in the early 80s, there was a widespread idea that the behaviour of superfluid helium was not exactly like that of a Bose-Einstein condensate. In other words, superfluid helium remained a puzzle to be solved.

From a qualitative point of view, an answer had already been given by Richard Feynman, who applied his original approach to the problem of the behaviour of

---

[9]D.M. Ceperley, B.J. Alder, *The calculation of the properties of metallic hydrogen using Monte Carlo*, Physica B **108**, 875, 1981; D.M. Ceperley, B.J. Alder, *Ground state of solid hydrogen at high pressure*, Physical Review B **36**, 2092, 1987.

quantum particles known as the path integral method. Developed in 1948, this was a quantum generalization of the action principle in classical mechanics. In practice, it replaced the classical single trajectory of a particle with a sum (i.e., an integral) over the infinite number of possible trajectories of a quantum particle, proposing a probability amplitude (or density) for these trajectories. Feynman exploited this approach to discuss the behaviour of liquid helium in a series of papers published in 1953.[10]

Ceperley had very clear ideas: «In my opinion, the path integral theory of Feynman (1953) explained the origin of superfluidity but this was not accepted by many people. Feynman showed that you can map a quantum many-body system onto a classical system of "polymers". The puzzle to explain is why, if the transition temperature of an ideal Bose condensate at the helium density occurs at about 3.1 K, the observed transition temperature in $^4$He is 2.17 K. How is it that you can have a strong helium-helium interaction and only shift in Tc by less than a factor of 50%? The interaction doesn't seem to do enough. Feynman explained that helium is a low density liquid. The atoms can still move around and exchange. If you think about it in terms of non-interacting single particle states you can't understand BEC, but if you think about them in terms of polymers, you can. While in BEC you have almost 100% of the atoms in the condensate, in helium it is less than 10%. How can the condensed atoms stay out of the way of the non-condensed atoms? Feynman's intuition», Ceperley goes on, «wasn't based on any calculation, it was an intuitive picture with "hand-waving" approximations. Pollock and I were familiar with Feynman's theory; there's a chapter on it in his book on statistical mechanics. But the imaginary time path integral methods were not commonly used computationally at that time since they required simulations, and much more difficult ones than classical simulations».

In fact, they were not the only ones who used the path integral method. In the early 80's, the idea that Feynman's theory might offer a promising new terrain for developing quantum simulation started to spread around. At the root of this interest, there were two particular features of this approach. The first was the formal analogy, already introduced qualitatively by Feynman, between virtual paths and polymers: the theorists working on an analytical approach to the problem explicitly introduced the term "isomorphism" to indicate this analogy.[11] This circumstance opened the prospect of bringing the problem back to a classical equivalent: to simulate a quantum system with a path integral method might simply mean to simulate the behaviour of a polymer chain. And polymers were by then a well-established field of molecular simulation.

Above all, what made this approach interesting was that the path integral method could treat the behaviour of the system at finite temperature, unlike all the other Monte Carlo versions, which were limited to handling the ground state at absolute

---

[10]R.P. Feynman, *The λ-Transition in Liquid Helium*, Physical Review **90**, 1116, 1953; R.P. Feynman, *Atomic Theory of the λ Transition in Helium*, Physical Review **91**, 1291, 1953; R.P. Feynman, *Atomic Theory of Liquid Helium*, Physical Review **91**, 1301, 1953.

[11]D. Chandler, P.G. Wolynes, *Exploiting the isomorphism between quantum theory and classical statistical mechanics of polyatomic fluids*, Journal of Chemical Physics **74**, 4078, 1981.

zero. Simulations of the behaviour of helium realized in previous years, like those by McMillan in 1965, or by Kalos, Verlet, and Levesque in 1974, could get results only for the ground state. With similar approaches, it was not possible to explain the existence of a superfluid transition at a finite temperature, $T_c$, but only superfluidity at $T = 0$. This is due to the fact that the existence of $T_c$ brings into play excited states.

Ceperley and Pollock decided to follow an original direction, simulating the behaviour of liquid helium, in particular $^4He$, making effective use of the potentialities of Feynman's approach. The experience with Lebowitz could finally be put to good use: «Here, the work I had done on polymers came in handy, since it had required that I dealt with topics such as dynamics, entanglement and cross-linking of long atom chains». Moreover, the two scientists explicitly took into consideration the quantum nature of the system, considering, in the calculation algorithm, the permutations connected to the property of the particles of being indistinguishable, as required by quantum statistics.

The work started when David and Roy read a paper published by John Barker at IBM in San Jose, where path integrals were used to study hard spheres.[12] Ceperley was impressed, and immediately organized a seminar in which, beside Pollock and Barker, Alder and Kalos were also present. The following discussion convinced Ceperley and Pollock that Barker had shown the right direction. Indeed, path integrals took into consideration finite temperatures, and could handle the problem of superfluid transition, whereas Kalos's method was limited to the ground state at absolute zero.

Without going into the technical details, the method which Ceperley and Pollock decided to follow allowed them to avoid a series of possible errors, such as those deriving from an inappropriate choice of the trial wave function, which was not necessary when working with path integrals. However, there was another side to it: they had to carefully evaluate how to sample the space, by inventing a new procedure which also took into account the permutations. Thanks to the isomorphism between Feynman's paths and polymers, Ceperley's experience with the latter came in particularly handy. Among other things, David was truly brilliant in coming up with techniques that could significantly reduce calculation times, by shortening the number of steps required to get a correct sampling of paths/polymers. Apart from these difficulties, there was no doubt: simulation with path integrals turned out to be more effective and more powerful.

In July 1983, Ceperley and Kalos met with Rahman in Trieste. During the meeting the problem was raised of the best possible way to implement the path integral technique, whether Monte Carlo or molecular dynamics. According to Ceperley, the answer was clear. It is true that, as a general rule, molecular dynamics can calculate both the static and the dynamic properties at the same time of a system. However, this justification does not apply to quantum simulation, since "displacements" in the space of Feynman's virtual paths had nothing to do with the effective dynamics of the quantum particles.

---

[12]J. Barker, *A quantum-statistical Monte Carlo method; path integrals with boundary conditions*, Journal of Chemical Physics **70**, 2914, 1979.

Ceperley had the opportunity to discuss a few aspects of his method directly with Feynman: «It was during that period (Jan. 1983) that I met Richard Feynman and had a chance to describe to him a computer implementation of his path integral theory of superfluid helium». The occasion for this discussion was the meeting organized by Werner Erhard, director and founder of Erhard Seminar Training, a "movement for human potential", whose main seat was in San Francisco (established in 1971, it was closed down the year after this meeting, in 1984). Erhard had gathered a group of thirty people to talk about the frontiers of science. Among them were Feynman, Kalos, and Ceperley. In that period, Feynman was particularly interested in the possibilities offered by the development of quantum computing, and had recently published a critical essay on the perspectives in this field.[13] Ceperley could therefore discuss with him the distribution of permutations in the superfluid transition.

Ceperley and Pollock presented—together with Alder—the preliminary results of their work in 1982, in a meeting whose proceedings were published later on. The description of the algorithm, as well as the detailed explanation of the method and of its limitations, appeared in a long paper in 1984, signed only by David and Roy, followed by a short communication in 1986, which summed up the results obtained on the properties of low temperature liquid helium.[14] At this point in their research, Ceperley and Pollock could announce that the "unusual properties of liquid He at low temperature", attributed to Bose-Einstein condensation, whose determination from first principles had so far been prevented by the strong pair interaction between atoms, were now explained and calculated: «In this Letter we present a Monte Carlo discretized path-integral computation of the density matrix for liquid $^4$He for temperatures spanning this transition [the Bose-Einstein condensation] which reproduces many of the experimental results and is in principle capable of arbitrary accuracy». Ceperley would later publish a review paper on the method, results, and details of the algorithm.[15]

In time, Ceperley expressed a very clear opinion on the relevance of these papers: «This is one of the times that looking at the simulation led to a much deeper theoretical result and something that has important implications for simulations. With the calculation of the superfluid density, we had explained many of the unusual features of superfluid helium starting from the interatomic potential. No adjustable parameters or uncontrolled approximations were needed. I think the Path Integral Monte

---

[13]R.P. Feynman, *Simulating physics with computers*, International Journal of Theoretical Physics **21**, 467, 1982.

[14]B.J. Alder, D.M. Ceperley, E.L. Pollock, *Computer simulation of phase transitions in classical and quantum systems*, in *Proceedings of the International Symposium on Quantum Chemistry, Theory of Condensed Matter, and Propagator Methods in the Quantum Theory of Matter*, International Journal of Quantum Chemistry **22** (S16), 49, 1982; E.L. Pollock, D.M. Ceperley, *Simulation of quantum many-body systems by path integral methods*, Physical Review B **30**, 2555, 1984; D.M. Ceperley, E.L. Pollock, *Path-integral computation of the low-temperature properties of liquid $^4$He*, Physical Review Letters **56**, 351, 1986.

[15]D.M. Ceperley, *Path integrals in the theory of condensed helium*, Review of Modern Physics **67**, 279, 1995.

Carlo has had an even larger impact on theoretical physics than the zero temperature quantum Monte Carlo calculations, since it is more closely related to classical statistical mechanics».

This was also why Ceperley was surprised when someone told him that his algorithm, based upon path integrals, was "baroque", i.e., too complicated. Ceperley was struck by this criticism. Later on, his thoughts on the matter were particularly enlightening: «I was somewhat surprised by this criticism and after reflection did not think the algorithm was more complicated than it needed to be. Some think the simulation of the Ising model represents the ideal simulation, the simplest possible model, and a very simple Monte Carlo algorithm to determine its properties. The goal of fundamental research [according to this viewpoint] should be to find such simple models and algorithms to calculate universal properties such as critical exponents. However, I had a different goal: can we go from the interatomic potential and calculate the experimental properties without adjustments. This required solving a number of problems». In sum, Ceperley suggested, simulation is precious not so much because it allows us to represent the universal features of a few particularly surprising laws (the properties of critical points in second-order phase transitions), as because it enables us to enter into the complexities of reality, which is ruled by the universal laws of physics, and, at least in principle, can be deduced from those laws.

After developing the method to simulate the superfluid phase in liquid helium, the two scientists went their separate ways. While Pollock started to calculate the superfluid fraction in the liquid, Ceperley took a sabbatical. This was 1985. Ceperley, after seven years at the University of California, obtained a sabbatical and decided to spend it in Italy. In the month of November 1984, during a symposium in honour of Rahman, David had met Gianni Jacucci, who asked him to come and work in Trento; in Povo, to be precise. It was a really quiet place, dominated by the Dolomites. These were the first years of the World Wide Web, and Ceperley was fascinated by the opportunity to send his codes to Livermore and its computers without material supports and a long wait. That was just another kind of life. However, beside his connections with the USA, thanks to Jacucci, Ceperley gained access to the Cray computer at Cineca, in Bologna.

Ceperley stayed in Trento from the fall of 1985 until the spring of 1986. He spent these months studying the magnetic phase diagram of solid $^3$He using the same procedure with which he had studied liquid $^4$He. He was encouraged to take up these studies by conversations he had had in February 1985 with J.M. Delrieu, M. Roger, and J.H. Hetherington on the occasion of a conference, and with Michael Fisher at the Aspen Institute for Physics.

The $^3$He atom is a fermion, that is, it has half-integer spin and obeys Pauli's exclusion principle. The physics of magnetic phases in crystals is normally described using effective lattice Hamiltonians, such as, for instance, the Ising (classical) and/or Heisenberg (quantum) Hamiltonian. The parameters of these effective Hamiltonians are normally treated as phenomenological parameters, to reproduce experimental data. In the case of $^3$He, the phenomenology of the magnetic phase is more complex. Its description requires the introduction of a lattice Hamiltonian with many-body

(3- or 4-body) terms. The formulation of these spin coupling coefficients may be given in terms of the path integrals of the atomistic model. However, in this formula, a direct calculation faces the difficulty of simulating rare events. Jacucci knew this aspect of the problem very well, and introduced Ceperley to Bennett's method for evaluating free energy differences. As Ceperley remarked: «That's a nice feature of path integrals; since they map onto classical mechanics, you can use a technique that has been developed for classical mechanics for quantum calculations» . Thus Ceperley thought of using, for this calculation, the same path integrals which had been used to calculate $^4$He properties, complemented by Bennett's method. The results he obtained, starting only from the interaction potential among helium atoms, showed that the calculated couplings had the expected values required to describe experimental results. The paper which summed up the simulation on $^3$He was published in 1987.[16]

Coming back to the USA in 1986, Ceperley continued to discuss with Pollock about path integrals and fermions. Within the next few years, he obtained quite interesting new results, such as the generalization of the so-called fixed-node procedure to the treatment of fermions with path integrals.[17] Meanwhile, in 1987, Ceperley accepted a permanent position at the University of Illinois in Urbana-Champaign. «After I moved to Illinois, my research style changed from working primarily by myself to working with students, post-docs and new collaborators». However, in the meantime, quantum simulation had established itself and had become an important area within the expanding territory of molecular simulation.

## 7.3 Kurt Binder: Scaling Laws and Universality

Towards the end of the 60s, a particularly lively field in the theoretical physics of condensed matter states concerned critical phenomena. This term included a category of phenomena in which physicists recorded a peculiar behaviour of observables, such as free energy, magnetization, magnetic susceptibility, correlation length, for particular (critical) values, of the system's thermodynamic variables.

The notion of critical point dates back to the end of the nineteenth century, when it was introduced in thermodynamics to define the particular conditions of maximum (critical) temperature and pressure under which the gaseous and liquid phases of a substance may still coexist. Before long, the notion of criticality was extended, in general, to other kinds of phase transitions, such as the transition between para- and ferromagnetic phases, studied by Pierre Curie, who determined the critical temperature at which the transition takes place. As a general rule, these phenomena represent the collective behaviour of a large number of strongly correlated components.

---

[16]D.M. Ceperley, G. Jacucci, *Calculation of exchange frequencies in bcc $^3$He with the path integral Monte Carlo method*, Physical Review Letters **58**, 1648, 1987.

[17]D. Ceperley, *Fermion nodes*, Journal of Statistical Physics **63**, 1237, 1991; D. Ceperley, *Path-integral calculations of normal liquid $^3$He*, Physical Review Letters **69**, 331, 1992.

The theoretical models developed in the first half of the twentieth century (such as Weiss' mean field theory for ferromagnetism, or the general theory of phase transitions, proposed by Landau) could not account quantitatively for the general features of the critical phenomena emerging from experimental data gathered with the techniques available in the 60s. For an advancement in theoretical understanding, new concepts were introduced, such as the characterization of the behaviour of a thermodynamic system through critical exponents, appearing in the power laws describing the behaviour of those physical quantities which develop singularities in the transition region; or the idea of scale invariance, and its connection with the correlation length tending to infinity near the critical point (this property becomes macroscopically visible as critical opalescence when, at the critical point, the liquid mixes up continuously with the gas).

Many of these ideas took shape on the basis of results obtained by Lars Onsager in a well-known paper of 1944 on Ising's model, which is the simplest schematization for a magnetic system[18], i.e., an array of spins ("magnetic needles") placed at the corners of a two-dimensional lattice, which can only be oriented either parallel or anti-parallel to a given direction, and only interact with their immediate neighbours. Onsager managed to calculate the statistical properties of the model exactly at all temperatures, thus highlighting the existence of a ferromagnetic phase transition. Notwithstanding the simplicity of this model, which could hardly be used to represent real systems, this work was of primary importance, since it showed that the possible existence of a second order phase transition is contained in the first principles of statistical mechanics.

By the end of the 60s, Onsager's was the only exact result yet achieved in the statistical mechanics of critical phenomena. Notwithstanding the appearance of new ideas to characterize properties near the transition, analytical solutions remained out of reach. Above all, while in areas such as the physical chemistry of liquids simulation was by now accepted as a research tool, the peculiar difficulties exhibited by critical phenomena made scientists think that it would be impossible to approach such problems from a computational standpoint. That would mean studying situations characterized by peculiar singularities in the values of physical variables, with correlation lengths going to infinity, thus necessarily involving the simulation of systems with a prohibitively high number of components; and the undertaking was made even more difficult by the circumstance that, in the critical region, thermodynamic properties show strong fluctuations which would tend to destroy the convergence of the average values to be computed.

In sum, while in statistical mechanics the study of critical phenomena was imposing itself as the main problem of research for theoretical physics, there was apparently no chance of any contribution on the part of simulation. Starting in the 70s, Kurt Binder took it upon himself to prove the opposite, developing suitable computational schemes which enabled him to integrate the Monte Carlo method in the theory of critical phenomena.

---

[18]L. Onsager, *Crystal statistics. I. A two-dimensional model with an order-disorder transition,* Physical Review **65**, 117, 1944.

We mentioned Kurt Binder earlier on when we said that, between the end of the 60s and the early 70s, computer simulation was spreading, even from a geographical point of view, from the United States towards Europe. In particular, Binder contributed to the diffusion in Europe of simulation using the Monte Carlo method. We left him in 1969, when he had just got his PhD in physics. It is worthwhile to recall the studies which led him to get this degree.

As a compromise solution between his own aspirations towards theoretical physics and his family's expectations towards a practical career, Binder enrolled at the Technische Universität, rather than at the University of Vienna. However, he was not enthralled by the kind of physics he studied at the polytechnic, whereas he found the work at the "Atominstitut", located next to Vienna's amusement park, Prater, very interesting. Indeed, there was a research nuclear reactor. It was managed by Gustav Ortner, but it was his assistant, Helmut Rauch, who was the real scientific mind at the centre. Actually it was Rauch, a specialist in neutron interferometry, who assigned Binder the task he should fulfil to get his Master's degree. The Prater reactor produced a weak neutron beam, and Kurt had to interpret some intensity data of these neutrons emerging from the beam guiding tubes. Binder's explanations were close to the target. «This success», he would later remember, «gave me an immediate reputation as a theorist able to explain experimental results».[19]

The task which Rauch assigned Binder for his PhD dissertation, after appreciating his degree work, required an explanation of the experimental data which had been obtained when a sample of ferromagnetic material was hit by the neutron beam. In Rauch's experiment, the sample temperature was steadily varied until it reached the Curie temperature, i.e., the temperature at which the material would undergo the transition from ferromagnetic to paramagnetic. In close proximity to the critical temperature, the data showed a strong anomaly in the density of the transmitted neutron beam, and Binder was asked to provide a quantitative interpretation of this anomaly. This was Binder's first encounter with the world of critical phenomena, which affected most of his future scientific activity.

At the Polytechnic Institute where he studied, Binder had no theoretical physicist to relate to on the topics connected with critical phenomena; also, he had no relationships with the University, which actually had a school of theoretical physics centred around Walther Thirring. Therefore, left to himself, Binder started to familiarize himself with the literature on the topic, and discovered he had come across a problem around which, in those very years, an intense theoretical activity was developing. Leo Kadanoff and collaborators had just published a paper on critical exponents, and Michael Fisher had just published his contributions.[20] The idea of the universality of critical phenomena was by then starting to gain acceptance. Soon Binder was convinced that it was more interesting and more promising to get involved in models

---

[19]This quote and the following come from the interview given by Binder on October 2, 2018 at the University of Mainz, never published, and sent to the authors by the interviewer, M. Mareschal.

[20]L. Kadanoff et al., *Static phenomena near critical points: Theory and experiments*, Review of Modern Physics **39**, 395, 1967; M.E. Fisher, *The theory of equilibrium critical phenomena*, Reports on Progress in Physics **30**, 615, 1967.

of magnetic systems and critical phenomena, than to run after improbable solutions to the problem of neutron anomalies: «I was then in trouble to explain to my thesis advisor that I was willing to study critical phenomena in ferromagnets, but that he should not expect that I would ever be able to explain the origin of the anomaly in his measurements of the transmitted beam». Rauch, who for all practical purposes played the role of advisor, even if he could not be officially labelled as such because he did not have the required academic status, accepted Binder's idea.

Thus Binder devoted himself fully to the study of critical phenomena in magnetism, mastering the formal techniques needed to work on the Ising and Heisenberg models (the latter is a more refined variant of Ising's first model, in which the magnetic moments of the atoms can be oriented in any direction). He worked on his own, without theoretical guidance, just when this field was undergoing a rapid evolution and new ideas were taking shape. The first results of this apprenticeship were weighty calculations of high-temperature serial expansions for Heisenberg's ferromagnetic model, up to higher and higher orders, which would soon turn out to be basically useless.

A new perspective opened up for Binder when it was suggested that he should take part in a series of seminars on the physics of liquids, organized by Peter Weinzierl, an experimental physicist. On this occasion, the young Binder heard for the first time about computer simulation with the Monte Carlo method. Binder discovered that, with this method, one could calculate correlation functions between particles within a fluid with a precision one could never reach with any analytical method.

That was a real epiphany. Binder started studying the literature on the Monte Carlo method, and started thinking about how he could apply it to his own problem. Indeed, he wanted to apply it to the model proposed by Ernst Ising, later extended by Werner Heisenberg and others, to explain the transition of a material from ferromagnetic to paramagnetic, once above the Curie temperature.

That was an original idea. Of course, both Monte Carlo and molecular dynamics methods were by then largely accepted techniques within the community of chemical physicists working in statistical mechanics, and they had given good results in the study of liquids. However, the field of critical phenomena was a different matter. As a general rule, these phenomena were barely understood, mainly because no way had been found to explain "critical" transitions, such as the one, at a given temperature, from ferromagnetic to paramagnetic, which take place with singular behaviour of the various observables (susceptibility, specific heat, etc.). The available theories did not exactly locate the critical temperature which may be observed in nature. Moreover, they did not provide the exact singular behaviour of the material above and below the critical temperature. And in addition, computer simulation of critical phenomena appeared to most physicists to be a hopeless enterprise. Usually, in close proximity of the critical point, correlations extended to extremely long scales; the necessary finiteness of the models one could treat, and the limited calculation power, would not allow scientists to see the typical features of the phenomena under study. In the 60s, the first few attempts to produce results on the Ising model with the Monte Carlo method hardly managed to simulate the behaviour of a cubic lattice with 64 ($4 \times 4 \times 4$) spins, obtaining wholly inconsistent results, and those theorists active in the field were convinced that simulation was an impossible route for understanding critical phenomena.

At that point, Kurt Binder developed his idea: why not apply computer simulation with the Monte Carlo method to ferromagnetic systems of growing dimension, and then check if and when there is a point of convergence? If this was successful, he thought, he would finally be able to explain the phenomenon of transition from ferromagnetic to paramagnetic under real conditions. That was an idea that had already been put forward (Alder had referred to it in 1952), which Binder would later develop in an impressive series of contributions over the following years, leading to the method of finite size scaling. «I felt that one could try for larger systems and that one could try doing different system sizes. This approach of varying the size was already in my first paper on the subject. I had the intuitive idea that one should try different system sizes and try to extrapolate to a large system. I knew that 64 Ising spins was far from the thermodynamic limit, and there was some word of caution in my papers, but I was the one who started varying the size to see how it converges and then study the effect in order to extrapolate».

Ferromagnetism is that phenomenon in which all the atoms of a material spontaneously align their spins along the same direction, just like the needles of a collection of compasses, all pointing towards the north. This is a particular case of the so-called spontaneous symmetry breaking; in this particular case, the symmetry consists in the isotropy of directions in space, which is broken by the alignment of spins along a single preferred direction. By following a procedure used for the study of liquids with the Monte Carlo method, Binder found the way to simulate on a computer the behaviour of ferromagnetic materials at a truly microscopic level.

This method was a little convoluted, and required a large amount of computing. Binder had to work by night, when he got six hours of machine time, but, in any case, the data he obtained were few. Fortunately, they were convergent. Thus he achieved significant results, which were proposed in two papers. One of these was signed with Helmut Rauch who, we should recall, was an experimentalist. The paper was sent in May 1968 and concerned the calculation of the spin correlation function in a ferromagnet using the Monte Carlo method.[21] Correlations were calculated at six different temperatures, both above and below the Curie temperature.

This paper consisted, basically, in the computational part of Binder's PhD dissertation. Following university rules, the dissertation was read and commented upon also by the professor of theoretical physics, Otto Hittmair. Binder's work contained the results obtained using the Monte Carlo method. As mentioned earlier on, at the end of the 60s computer simulation was still looked upon with haughtiness by most theoretical physicists. At best, it was considered as a computer experiment. Nobody was likely to consider it as a new way to make physical theory. Notwithstanding all this, Binder's dissertation received an excellent evaluation, since it included a substantial part devoted to purely analytical developments.

---

[21] K. Binder, H. Rauch, *Calculation of spin-correlation functions in a ferromagnet with a Monte Carlo method*, Physics Letters A **27**, 247, 1968.

Binder also pointed out the limitations of his approach. In the last paragraph of his first paper, he wrote that the number of examined lattice points was too small, due to computing limitations. It was too small to exactly determine the Curie temperature. However, he succeeded in establishing an interval of temperature in which the phase transition would take place. To make a comparison with the experimental data on macroscopic systems, he concluded, one would need more magnetic moments on the sites of the studied lattice, and therefore a higher computing power.

Once he had obtained his PhD, the young physicist started imagining his own future. He thought he hardly stood a chance for a job at the University of Vienna. He had a theoretical dissertation which had not been supported by the professor of theoretical physics. And, though he had certainly achieved relevant results, the Monte Carlo method was basically unknown in Austria. Who would give him a job under such conditions?

Probably nobody in Vienna, but certainly a lot of people in Germany. On the other hand, Kurt wanted to leave his country, so as not to put a huge financial burden upon his family. He sent his CV to a good number of German research centres, and received more than one positive answer. He chose the offer from Heinz Maier-Leibnitz, a well-known experimental physicist who had formed a group in Munich to cultivate reactor physics. This group needed a theorist, and Leibnitz thought Binder would be the right person.

In 1969, Binder moved from Vienna to Munich. Soon afterwards, he heard about a summer school to be held in Varenna, Italy, during the summer of 1970, focussing upon critical phenomena. Binder asked and obtained the permission to go there. Kurt Binder left Munich for Varenna together with Dietrich Stauffer, a postdoc working in the group of Wilhelm Brenig, one of the most influential theoretical physicists in Germany.

The summer school in Varenna was a crucial moment in Binder's scientific career. Indeed, on this occasion he came directly in touch with the main specialists on phase transitions; among the school lecturers were Leo Kadanoff, Michael Fisher, Bob Griffiths, and Pierre Hohenberg. There was also Gene Stanley, who was finishing off a monograph on phase transitions and critical phenomena, which would become the main, and the most authoritative, reference on this topic (he published it the following year).[22] «It should be recalled that understanding critical phenomena then still was an outstanding challenge and heavily debated; although some lectures on mathematical aspects of the renormalization group were given by Giovanni Jona-Lasinio, it remained obscure – at least for me – how this ever could help to understand critical phenomena. Only a year later the papers by Kenneth G. Wilson provided a breakthrough in understanding».[23] Binder realized that, in order to make progress with his research programme, it would not be enough to rely on the growth of

---

[22]H.E. Stanley, *Introduction to Phase Transitions and Critical Phenomena*, Clarendon Press, Oxford, 1971.

[23]K. Binder, *From the Lake of Como to surface critical phenomena*, in J. Brujic, A Grosberg (eds.), *Memories of Pierre Hohenberg*, New York 2018, pp. 30–34, cit. p. 31. Wilson's papers on the renormalization group are: K.G. Wilson, *Renormalization group and critical phenomena. I. Renormalization group and the Kadanoff scaling picture*, Physical Review B **4**, 3174, 1971; K.G.

computer calculation power. He would also need to develop new and creative ideas, and acquire a certain theoretical skill which would enable him to discuss on an equal footing with the people he had met in Varenna.

The 1970 Varenna summer school on critical phenomena. Kurt Binder is the second from right in the third row. (Reproduced with permission from: *Proceedings of the International School of Physics "Enrico Fermi"*, Course LI, "Critical Phenomena", edited by M.S. Green ©1970 Società Italiana di Fisica)

Among these, Pierre Hohenberg was particularly interested in Binder's work, when Binder explained his research in one of the school's afternoon sessions reserved for young scientists. Pierre Hohenberg was one of the main theorists of critical phenomena and phase-transitions at Bell Laboratories, in New Jersey. Meeting Hohenberg was important for two reasons: first of all, Binder was encouraged by one of the best specialists in the field; secondly, Hohenberg told him that he wanted to spend a period in Munich, and declared himself ready to collaborate. Indeed, towards the end of 1971 Hohenberg took a sabbatical, and spent it in part (six months) in Munich, where he met Binder once again. Better, he actually met Binder alone, since, during those six months in Munich, Hohenberg worked with Binder alone. The aim of their studies was surface critical phenomena in magnets.

As soon as Hohenberg arrived in Munich, Binder showed him the latest results he had obtained on the critical behaviour of the surface susceptibility of an Ising ferromagnet, by systematic extrapolations of the calculations made with high-temperature series expansions. From his results, he could extrapolate the critical point, that is, the point at which a phase transition occurred. However, the exponents obtained by Binder for the critical point on the material surface were quite different from those for bulk matter. That was an effect for which Binder had no explanation, but Hohenberg could explain this, so they started working together, calculating the surface critical exponents. «We started then interacting about this. What are the independent exponents here, is the correlation length in the surface different from what it is in the bulk? Pierre discovered then that there was a need to introduce actually three

Wilson, *Renormalization group and critical phenomena. II. Phase-space cell analysis of critical behavior*, Physical Review B **4**, 3184, 1971.

susceptibilities: because you can introduce a field which acts in the bulk, and you can also introduce a field that acts on the surface, they are different [...] This was very plausible, of course, and the task was to get all those susceptibilities from the series expansions. And then to figure out whether the exponents were independent exponents, or related to each other by scaling laws. We needed a scaling description for these phenomena, there was a zoo of new critical exponents: one could consider the decay of the spin correlation function in the bulk, at the surface, in the direction normal to the surface or parallel to the surface, etc. And how these things are related to each other. We were discovering new scaling laws every day before going to lunch! This happened several times, it was a very pleasant interaction. Soon, we had discovered the new scaling relations for surface critical phenomena!».

The result of this work was a huge 100 page manuscript. The poor secretary who had to type it was shocked. In Munich, theoretical physicists usually wrote papers a few pages long, rather than books. Typing a hundred pages certainly was no pleasant outlook. Moreover, the work made by this team of theoretical physicists was full of formulae: a real nightmare. However, the secretary scrupulously completed her task, and the paper was soon published in 27 pages of the *Physical Review*.[24] It was a purely analytical work, with no mention of the computational approach. But that is what Binder was looking for. He wanted to establish himself in his own right in the theoretical condensed matter community, showing against any remaining doubts that simulation was, to all intents and purposes, the key tool for attacking and solving the crucial open problems in the field.

## 7.4  Criticality Everywhere: From Ferromagnetism to Polymers

At the beginning of this period, a novelty was introduced by Wilson's work on the renormalization group. Picking up on a technique developed about twenty years earlier to solve the problems of the divergences in the calculations of quantum field theory, Wilson built a powerful formalism which allowed one to treat similar problems in the theory of critical phenomena, embedding the concepts which had been developed in previous years in a mathematically sophisticated formulation. Such concepts included the scaling laws and the properties of universality arising when, in different systems, one gets closer to the critical region.

Binder followed these developments and transformed them into ever better performing programmes, adapted to a wide range of critical situations. The huge amount of work done in the years following the Varenna school and the meeting with Hohenberg produced a long series of papers where analytical contributions, with which Binder showed that he had become an authoritative theorist in the field and gained scientific and academic credibility in the community of mainstream theoretical physics,

---

[24]K. Binder, P.C. Hohenberg, *Phase transitions and static spin correlations in Ising models with free surfaces*, Physical Review B **6**, 3461, 1972.

alternated with increasingly effective codes and algorithms which could be used to submit the main problems of critical phenomena to simulation.

In the month of April 1972, Binder moved from Munich to Zurich, with a one-year post-doc grant, at the IBM research laboratory, where he could take full advantage of the exceptional potentialities offered by the lab computers, with an "almost unlimited" quantity of computing resources and machine time. Thanks also to the excellent working conditions and the computers available at the IBM laboratory, the year spent in Zurich was particularly productive. In collaboration with Heiner Müller-Krumbhaar, a student from Munich, whose dissertation he was supervising off the record, he started using the Monte Carlo method to produce significant results on the critical behaviour of the Heisenberg model. After a few preliminary results, they published a paper in which they presented the data obtained through computer simulation on magnetization, susceptibility, and evolution of the magnetic field at the Curie temperature. These data allowed a direct evaluation of the critical exponents with an accuracy comparable to what could be achieved using analytical techniques of extrapolation, or from the renormalization group.[25]

The papers on the Heisenberg model were a generalization of those on the Ising model, following a line of research in which discrete lattice models provided the key to calculate the properties of critical phenomena. On the computation front, Binder started getting interested in the kinetic properties of the Monte Carlo method. Still in collaboration with Müller-Krumbhaar, he published a paper in which he showed that he could interpret the sampling of the Metropolis Monte Carlo method in terms of the kinetics of a stochastic process, introducing an arbitrary time scale to turn the transition probability into a transition probability per unit time.[26] In this way, the correlation between the microstates which were subsequently produced in the sampling procedure could be understood in terms of the correlation time of the master equation describing the process.

In the same period, Binder started getting interested in nucleation, i.e., the physical process by which clusters are formed and subsequently trigger a phase transition. Nucleation is a very general phenomenon which can be found in any kind of transition. Significantly, Binder only examined the problem for critical transitions, which posed the biggest theoretical challenge. Thus he started a collaboration, which would be resumed more than once in the course of time, with Dietrich Stauffer, a theorist whom Binder had introduced to the world of simulation. Starting in 1972, Binder and Stauffer, sometimes in collaboration with Müller-Krumbhaar, wrote a few papers on the clusters that trigger nucleation, and on their shape, focussing on the drop model created by Fisher and calculating its properties using the Monte Carlo technique.[27]

[25]K. Binder, H. Müller-Krumbhaar, *Monte Carlo calculation of the scaling equation of state for the classical Heisenberg ferromagnet*, Physical Review B **7**, 3297, 1973.

[26]H. Müller-Krumbhaar, K. Binder, *Dynamic properties of the Monte Carlo method in statistical mechanics*, Journal of Statistical Physics **8**, 1, 1973.

[27]K. Binder, D. Stauffer, *Monte Carlo study of the surface area of liquid droplets*, Journal of Statistical Physics **6**, 49, 1972; K. Binder, D. Stauffer, H. Müller-Krumbhaar, *Calculation of dynamic critical properties from a cluster reaction theory*, Physical Review B **10**, 3853, 1974; K. Binder, D.

Back from Zurich to Munich at the end of 1973, Binder obtained the qualification which enabled him to be a university lecturer. In 1974, he flew to the United States, spending six months at the Bell Laboratories, on the invitation of Pierre Hohenberg. Back in Germany, at the end of the summer he accepted an offer by the University of Saarbrücken, and became the leader of a new research group. Dietrich Stauffer, who had been mistreated in Munich, followed him. Binder stayed three years at Saarbrücken. Then, in 1977, he married and left Saarbrücken in order to take over the direction of a theoretical group at the Institute for Solid Research at the Jülich Research Centre. This centre housed the most powerful German computer, and is nowadays called the Jülich Supercomputing Centre (JSC). However, the tasks related to the management of a large research group took up too much of the time he wanted to devote to his research activity, and Binder essentially wanted to do science, rather than organize the work of other scientists. After six years as director of the group at Jülich, he decided to give up this role, and in 1983 he accepted a position as professor of theoretical physics at the University of Mainz, where he settled down for good.

In all these years, he continued the integration of critical phenomena in simulation techniques, with an impressive rate of production. In the decade going from the period in Zurich to his settling down in Mainz, Binder published, alone or with several collaborators, over a hundred and forty papers.

Let us resume the description of his research activity. As we mentioned above, in 1974, Binder spent six months at Bell Laboratories, on the invitation of Pierre Hohenberg. Prior to this stay, Binder and Hohenberg had finished their collaboration on surface critical phenomena and published a paper together.[28] This time, however, it was not a purely analytical paper. Rather, they presented results obtained in a computational way.

During his stay in the USA, thanks to Hohenberg, Binder came into contact with Mal Kalos, through whom he further explored the kinetics of the Monte Carlo method, and with Joel Lebowitz, who was very much interested in the potentialities of simulations and was struck by Binder's nucleation papers. Thus a collaboration sprang up among Binder, Kalos, and Lebowitz, and some of the results of this interaction were included in a review paper published in 1979.[29]

Still in the USA, at a seminar lecture by Phil Anderson, Binder got to know the problem of spin glasses. He started studying their behaviour with a computer, thus opening up another possible use of the Monte Carlo method which would keep him engaged for a long time.[30]

---

Stauffer, *Statistical theory of nucleation, condensation and coagulation*, Advances of Physics **25**, 343, 1976.

[28] K. Binder, P.C. Hohenberg, *Surface effects on magnetic phase transitions*, Physical Review B **9**, 2194, 1974.

[29] K. Binder, M.H. Kalos, J.L. Lebowitz, J. Marro, *Computer experiments on phase separation in binary alloys*, Advances in Colloid and Interface **10**, 173, 1979.

[30] K. Binder, K. Schröder, *Monte Carlo study of a two-dimensional Ising "spin glass"*, Solid State Communications **18**, 1361, 1976; K. Binder, K. Schröder, *Phase transitions of a nearest-neighbor Ising "spin glass"*, Physical Review B **14**, 2142, 1976.

In 1975 David Landau, a physicist from Georgia University, came to Saarbrücken for a sabbatical as "Alexander von Humboldt fellow" (a special position provided in the German University system for distinguished visiting professors from abroad). Landau had been educated as an experimental physicist, but after taking up his tenure track position in Georgia, he got involved in simulation as well. Binder had met Landau two years earlier at an international meeting on magnetism, and once again during his stay at Bell Laboratories. Together, Binder and Landau published a paper on the critical properties of the two-dimensional Heisenberg model. This was the first of a long series of papers published together.[31] They discussed how to make the most, for computational purposes, of the scaling laws characterizing critical phenomena, studied in the early 70s by Michael Fisher.

The singularities, typical of critical phenomena, manifest themselves in infinite systems. Obviously, real systems are finite, and a finite system cannot show a true singularity at a temperature other than zero. However, phase transitions do actually occur in nature, and it is therefore necessary to reconcile these two sides of the problem. The idea behind finite size scaling is that one may extrapolate the exact values of physical variables at the critical point by studying the behaviour of finite systems of different dimensions, taking into account the scaling laws which can be used to find their universality properties.

The implications of the finite size scaling hypothesis for the computer simulation of critical phenomena had long been the focus of Binder's interest. Making full use of this idea required an improved understanding of the underlying theoretical concepts (universality), and a computing power that was not available to him in earlier times. However, a few years later, during his stay in Zurich, the IBM computers enabled him to simulate the behaviour of a three-dimensional Ising lattice, which could extend up to $55 \times 55 \times 20$ elementary magnets (a "world record" at the time).[32] The paper which Binder published presenting these simulations gave encouraging indications, showing that finite size scaling provided the correct estimate of the critical temperature as a function of the correlation length, and pointing out that data tended to "collapse" when suitably rescaled variables were used.

This research occupied him for a long time, so much so that his most significant work would only be published a few years later, in 1981.[33] In this paper, Binder exactly assessed critical parameters by reconstructing, through simulation, the magnetization distribution function as a function of temperature in systems of different dimensions. More precisely, Binder calculated the particular role of a quantity derivable from this distribution function, namely the so-called "cumulant", related directly to the degree of disorder of the system. In the ideal case of an infinite system, this function is equal to zero above the critical temperature $T_c$ (when the system is in a completely disordered state) and takes on a definite value for temperatures lower than

---

[31]K. Binder, D.P. Landau, *Critical properties of the two-dimensional anisotropic Heisenberg model*, Physical Review B **13**, 1140, 1976.

[32]K. Binder, *Monte Carlo study of thin magnetic Ising films*, Thin Solid Films **20**, 367, 1974.

[33]K. Binder, *Finite size analysis of Ising model block distribution functions*, Zeitschrift für Physik B **43**, 119, 1981.

$T_c$ (this indicates the presence of a certain degree of correlation among the system components), with a sharp discontinuity, from this finite value to zero, at the critical temperature. On the other hand, for finite size systems, the function varies continuously (and faster when the system gets larger), from the known value at $T = 0$ down to zero for high temperatures. The simulations made by Binder showed that, however different the behaviour of this function may be according to the size of the system, all the curves end up at the same value at the critical temperature; the simulation demonstrated the universality properties of the critical point, and enabled him to assess the value of the critical temperature directly. In the same way, Binder showed that he could extract the values of critical exponents. This was a really stunning result; the Monte Carlo simulation, combined with finite-size scale analysis, beside enabling a stringent analysis of the critical behaviour of the finite Ising model, finally made it possible to calculate the critical parameters (i.e. temperature and critical exponents) without extrapolation.

By the late 70s, Binder was an established specialist in handling critical phenomena with the Monte Carlo method. In 1979, he edited a miscellaneous volume on the use of Monte Carlo methods in statistical physics, and this long remained a standard reference in the field.[34] On that occasion, Binder broadened his horizons beyond the pure domain of critical phenomena: apart from his own contributions, and those of his collaborators, such as Landau, Müller-Krumbhaar, and Stauffer, the volume contained two dense chapters on the simulation of classical fluids, and on quantum many-body problems, written respectively by Levesque, Weis, and Hansen, and by Ceperley and Kalos.

During the same period, Binder further extended the field of application of simulation.[35] Moreover, he exported the techniques developed for critical phenomena to problems such as polymer physics, apparently very distant but actually closely related.

In 1979, together with Artur Baumgärtner, he published his first paper on the application of Monte Carlo techniques to the study of polymers, and this was followed by several others.[36] He was encouraged by a few papers on this topic, published by Pierre Gilles de Gennes, who had stressed the analogies between the critical exponents of

---

[34] K. Binder (ed.), *Monte Carlo Methods in Statistical Physics*, Springer-Verlag 1979.

[35] His main motivation was to explain experimentally, in collaboration with David Landau, observed phenomena in various branches of solid state or surface physics. Another study, jointly with Joel Lebowitz and Mal Kalos, addressed long- and short-range order in alloys.

[36] A. Baumgärtner, K. Binder, *Monte Carlo studies on the freely jointed polymer chain with excluded volume interaction*, Journal of Chemical Physics **71**, 2541, 1979. The history of the joint development of theory and simulation of polymers has still to be written. However, it is worth noting that the theoretical physics of polymers was founded in the 1940s by Paul Flory, who later won the Nobel Prize for Chemistry for his work. The phase diagrams of polymer solutions and polymer blends were then modelled by what is now known as "Flory-Huggins theory", which became a kind of "gold standard" for experimentalists: P.J. Flory, *Thermodynamics of high polymer solutions*, Journal of Chemical Physics **9**, 660, 1941; M.L. Huggins, *Solutions of long chain compounds*, Journal of Chemical Physics 9, 440, 1941. Only in 1987 were the deficiencies of this mean-field theory shown by a Monte Carlo simulation: A. Sariban, K. Binder, *Critical properties of the Flory-Huggins lattice model of polymer mixtures*, Journal of Chemical Physics **86**, 5859, 1987.

phase transitions and the exponents describing how the dimension of polymeric coils scaled with the molecular weight. «Just as many people in magnetism had tried to improve on the Weiss molecular field theory, by making more sophisticated assumptions, and improving the theory more and more, and still could not describe the critical point behaviour [...] similarly in the polymer problem, people worked a lot on the standard ideas of Flory, calculating the properties of polymer chains with better and better approximations and still could not solve the self-avoiding walk problem, and describe its exponents. Then de Gennes showed how you could map mathematically this problem to a critical point problem. Immediately, polymer science became a respected field in theoretical physics, and was no longer considered something which was exclusively for chemists to worry about».[37]

De Gennes's work was purely analytical. However, by then, Binder had learnt to transfer the "noble" topics of theoretical physics to the terrain of simulation, so that he could include polymers, reinstated as objects worthy of attention for fundamental science, in the growing list of problems to be submitted to computation along his so-called "long walk in the land of criticality".[38]

Binder continued this long walk in the land of scaling and universality together with several collaborators, following a direction that kept him fairly isolated from the larger community of molecular simulators, at least until a period following the one discussed so far. His focus on critical phenomena derived from his ambition to be recognized as a fully theoretical physicist by the best-known specialists of statistical mechanics, some of whom were forced to abandon their basic scepticism towards simulation, which they typically considered as a pure numerical manipulation. On the one hand, Binder's intensive focus on a computer solution of Ising models, and its more or less complex by-products, kept him on the margin of the simulation world, but it also led him to a real breakthrough. Even the most reluctant "pure" analytical physicists could not be indifferent to his results; Binder forced his way into the field of "great" theoretical physics and showed that simulation could bring key innovations in fields other than the chemistry of liquids.

Therefore, by the mid-80s, even the two problem areas which at first seemed to escape the application of simulation, symbolically represented in the two figures introducing this chapter, were included within the range of problems that could be successfully solved using molecular simulation. Obviously, this was only the beginning of a process which is still going on today. However, thanks largely to the pioneering work of both Ceperley and Binder, the physics of quantum aggregates, and the physics of "criticality", i.e., scaling in critical phenomena and polymer physics (the

---

[37] Kurt Binder interview by D. Mac Kernan, SIMU Newsletter **3**, 19, 2001.

[38] M. Mareschal, *From Varenna (1971) to Como (1995): Kurt Binder's long walk in the land of criticality*, European Physics Journal H **44**, 161, 2019. It is worth noting that the attempts by simulators to study models of polymers by the numerical approach began with Metropolis MC (1953). Only two years later, the two Rosenbluths published their famous algorithm which opened the way to the study of restricted random walks on a lattice: M.N. Rosenbluth, A.W. Rosenbluth, *Monte Carlo calculation of the average extension of molecular chains*, Journal of Chemical Physics **23**, 356, 1955. Many researchers have since taken up the challenge of extending the Rosenbluth algorithm.

main problematic points in the theory of condensed matter) were by then embedded within the general programme of molecular simulation. And so the latter was shown to be a tool capable of handling the whole spectrum of behaviours of condensed matter in its various states of aggregation.

# Chapter 8
# A First Finishing Line and Some Provisional Conclusions

## 8.1 The First School of Molecular Dynamics: Varenna 1985

Around the mid-80s, molecular simulation had gone beyond the pioneering stage and had acquired the status of an established research area. It had been proved that a wide range of problems in the theoretical physics of the structure of matter could be faced and successfully solved in a computational way. As far as classical statistical mechanics was concerned, starting from the first pioneering attempts on simple and scarcely realistic models, such as the hard-sphere model, scientists went on to simulate clusters of simple atoms, and with a growing organizational complexity, more and more sophisticated molecules, until they addressed the problem of large biological molecules. The behaviour of simple fluids had been brought under control. Systems of constraints reducing the number of degrees of freedom of the systems under study, so as to start calculating dynamics on slower time scales, had been conceived, and efficient algorithms to implement them had been perfected. The invention of techniques aimed at extending the application of methods of molecular dynamics to generalized ensembles enabled physicists to submit the whole spectrum of thermodynamic transformations to simulation. The introduction of *ab initio* molecular dynamics opened the prospect of studying models in which chemical reactions made their first appearance. The exponential growth of computing power allowed the simulation of systems composed of so many particles (and for long enough times) that one could be sure to have reached the thermodynamic limit to an excellent approximation, thus obtaining fully reliable results where the analytical approach had previously failed. As for quantum systems, physicists could already simulate realistic models of boson systems at finite temperature, while the problem of simulating fermion systems was still open.

This process of growth had been accompanied by the informal rise of a lively international community, in a climate which was aptly defined by Daan Frenkel as "exhilarating"; «all the concepts of Statistical Mechanics, which until then had

© Springer Nature Switzerland AG 2020

G. Battimelli et al., *Computer Meets Theoretical Physics*, The Frontiers Collection,

https://doi.org/10.1007/978-3-030-39399-1_8

been abstract, had now acquired life. Thanks to computers, they could be calculated, therefore they were very real».[1]

A good indicator for assessing the degree of maturity of a discipline is represented by its capacity to create organizational structures, assuring the transmission and continuity of the knowledge which has been produced: from the phase of informal exchange among pioneers, one moves on to the stage of institutional stability, and the creation of a school tradition. The organization of dedicated schools is an exemplary moment in this growth. In our case, the event which made this process fully visible was the summer school organized in Varenna in the summer of 1985: the 97th course of the "Enrico Fermi" International Physics School of the Italian Physical Society was devoted to the "Simulation of statistical-mechanical systems with molecular dynamics".

The idea was born during one of the CECAM workshops in Paris, in the month of August 1983. Giovanni Ciccotti and Bill Hoover discussed the project of a summer school devoted to presenting the state-of-the-art of simulation in statistical mechanics. It seemed to them that the right time had come to organize a school (their goal being not so much, or not only, to discuss new results, but rather to train young researchers on the basis of acquired results) wholly devoted to simulation, avoiding the reproduction of yet another version of other scientific meetings, generically focussing upon problems of statistical mechanics in which simulation played a secondary role, as an auxiliary tool. They certainly did not want to organize yet another copy of the discussions upon fashionable topics of "pure" theoretical physics of critical phenomena. «We had in mind two recent IUPAP meetings, in Edmonton in 1980, and Edinburgh in 1983. It seemed to us that these international meetings on "Statistical Mechanics" were basically focussing upon developments of the Ising model, which Onsager had handled in the 40's [...] Our basic receipt for a school on the "Simulation of statistical-mechanic systems" was to carefully avoid the Ising model. In order to make this rule stricter, the term "Molecular Dynamics" was added later to the title».[2]

The project was discussed once again, and was well received, on the occasion of the meeting organized at the Argonne lab in the month of November 1984 to celebrate the twentieth anniversary of the publication of the fundamental work by Rahman on the dynamics of liquid argon. Unfortunately, Rahman was already experiencing severe symptoms of the illness that would lead to his death three years later, and he could therefore neither take part in the school, nor contribute in any way to the project. Among the founders of molecular dynamics, his absence was the only one really relevant at the school organized by Ciccotti and Hoover.

In the meantime, the Italian Physical Society (Società Italiana di Fisica or SIF) accepted the proposal of a course devoted to the developments of molecular dynamics. The school was held from July 23 to August 2, 1985, at the Villa Monastero, looking

---

[1]D. Frenkel, in *Recollections of CECAM – for Carl*, CECAM, Paris 1990.

[2]G. Ciccotti, W.G. Hoover, *Foreword*, in Id. (eds.), *Molecular-Dynamics Simulation of Statistical-Mechanical Systems*, Proceedings of the "Enrico Fermi" International Summer School of Physics, Course XCVII (Varenna, July 23 - August 2, 1985), North-Holland, Amsterdam 1986, p. XV.

out over Lake Como, the historical location of the SIF summer schools since 1954. According to official records, 25 teachers and 66 students met in Varenna. In fact, effective numbers were higher; there were a few observers passing through Villa Monastero, and the proceedings contained texts sent by lecturers who were not actually there. For instance, Andersen, who could not take part in the school for health reasons, sent his contribution for publication anyway. Organizational constraints prevented the organizers from exceeding a fixed number of participants, so that— among others—Jerry Percus, belonging to the "old" generation of analytical theorists, who had immediately looked sympathetically at simulation, was regretfully denied participation. In the end, 29 contributions were published in the volume collecting the lectures held during the course: an unprecedented record for the Varenna schools.

The group picture of the participants to the 1985 Varenna summer school on "Molecular-Dynamics Simulation of Statistical-Mechanical Systems". (Reproduced with permission from: *Proceedings of the International School of Physics "Enrico Fermi"*, Course XCVII, "Molecular-dynamics simulations of statistical-mechanical systems", edited by G. Ciccotti and W.G. Hoover, ©1985 Società Italiana di Fisica)

For the first time, "Molecular Dynamics" appeared explicitly in the title of such a school. In fact, the range of problems discussed during the school included also various aspects of Monte Carlo simulation. This was hardly surprising, in the light of the developments and of the frequent entanglements between the two complementary approaches to molecular simulation. Among the lecturers who explained the state-of-the-art in the discipline, there were many of the protagonists whose story we have followed in these pages. Three of them had been in Varenna in 1957, at the meeting which marked one of the starting points of this story: Bill Wood, who, in an introductory lecture, recalled the first stages of computer simulation in statistical mechanics; Berni Alder, who talked about the relationships between the microscopic approach and hydrodynamic description; and Andrè Bellemans, who closed the course with a general review of the topics that had been discussed, and gave his insights on the future of the discipline. Among the others, Herman Berendsen and Wilfred van

Gunsteren presented a review of the main algorithms used in equilibrium molecular dynamics, Jean-Pierre Hansen lectured on the simulation of Coulomb systems and the problem of long-range forces, and Daan Frenkel discussed the calculation of free energy and phase transitions. Berendsen gave a lecture on biological molecules, membranes, and proteins, whereas Brad Holian, Mike Klein, Gianni Jacucci, Ian McDonald, and Jean-Paul Ryckaert spoke about various aspects of the applications of molecular dynamics. As mentioned before, Hans Christian Andersen was not there, but sent his contribution anyway. In his lecture, Michele Parrinello explained the method he had just developed, together with Roberto Car, which had not yet been published (that was why he sent a completely different paper for publication in the proceedings). The Monte Carlo method, already presented in Frenkel's lectures, reappeared, in a more essential way, in the paper by David Ceperley, who, in the section devoted to quantum problems, discussed the recent results of QMC in the simulation of liquid helium. Carl Moser chaired the final roundtable on the prospects for the field and ongoing projects of dedicated computers for molecular dynamics simulation. Officially registered as "students", there were also John Valleau and Glenn Torrie. Indeed, many students of this course soon became, or were already well on the way to becoming, leading exponents of the discipline. We have mentioned some of them: Mauro Ferrario, Pierre Turq, and Michel Mareschal.

The Varenna school marked an important moment, in more than one respect. First of all, as we have said, it represented a significant step for the upgrade of molecular simulation to an institutional discipline, and, as a consequence, for its recognition as such on the part of the scientific community. Since a structure of international prestige, such as the SIF school, accepted to devote one of its courses to a topic that had up until then been ignored, a clear signal was sent out that this was no longer a minor research area, or at most an ancillary sector in comparison with truly significant research, but rather a discipline competing on an equal footing with contiguous research areas. Molecular simulation maintained its own autonomy and could count on an identifiable reference community.

Secondly, the school was held at a time which was particularly rich in important developments. We have seen how, during the 80s, a wide range of new results was produced, opening up several new perspectives to simulation. Many of these results were due to the school lecturers, who directly reported the results of their research, either in progress or just completed, in Varenna. Alder and Alley's review of *Generalized Hydrodynamics* was published in 1984, the revision of the idea of Nosé's thermostat on the part of Hoover had just been published, and the paper by Frenkel and Ladd on the Monte Carlo method for calculating the free energy of arbitrary solids was also published in 1984. The paper by Car and Parrinello was in press during the Varenna school, and in the summer of 1985 the papers on path integrals by Ceperley and Pollock were published. Still, just before and soon after 1985, the CECAM workshops took place, in which the constraint techniques leading to *Blue Moon* were developed, and various contributions to the problem of extending simulation techniques to generalized ensembles had recently been published. Varenna acted therefore as an effective resonance box for the novelties coming up in this research area, in which frontier problems of molecular simulation were discussed and presented to a new generation of researchers.

Indeed, and this is further proof of the relevance of the event, with the Varenna school a new generation appeared on the scene, which, although following the same research tradition, made a real difference from the founding fathers of simulation. The pioneers, who were the school lecturers, approached simulation (in a few cases it would be more accurate to say that they invented simulation) from directions that were often quite different from one another. At a scientific level, physicists or chemists, whether theoreticians or experimentalists by training, discovered simulation—and contributed to its growth—when they were in a more or less advanced stage of their intellectual life and career. When they gave life to this school, they gave shape to a community of researchers who were, so to speak, imprinted with simulation, and therefore ended up becoming unprecedented professionals, who—in a sense— redefined the traditional boundaries among disciplines.

In his closing remarks at Varenna, although he admitted the persistence of a "black hole", constituted by the absence of an effective method of molecular dynamics for quantum systems, Andrè Bellemans happily stated that "the situation is far from being frustrating".[3] In fact, the situation was excellent. Bellemans reviewed some of the recent steps forward which had been presented during the course: the techniques to "freeze" some degrees of freedom irrelevant for simulation in polyatomic molecules, the capacity to reconcile the existence of long-range forces with the periodic boundary conditions of the finite cells used in simulation, the possibility of relaxing the conditions of both finite volume and form in those same cells, the various recipes for extending the technique of molecular dynamics to ensembles different from the microcanonical one, and the developments of non-equilibrium molecular dynamics.

Bellemans remarked that these results confirmed that simulation had effectively satisfied three conditions: choosing important questions generated by either theory or experiment and giving significant answers to both theoretical and experimental physicists; giving information at an atomic level which could not be obtained in an experimental way, through the simulation of sophisticated models; finally, discovering new facts, and therefore posing stimulating questions to both theoretical and experimental physicists. By achieving these goals a sort of "quality label" was guaranteed which was not, on its own, automatically assured by the fact that the number of published papers was growing steadily. «Should we look at publications concerned with static and dynamic properties of liquids, then simulations have certainly reached a volume comparable to both theoretical and experimental works». However, Bellemans went on: «Volume is by no means a good criterion of respectability among the scientific community and the growing recourse to simulations (and numerical work in general), favoured by increasing computer facilities, has been sometimes criticized as endangering 'creativity of the mind'».

---

[3] A. Bellemans, *Synopsis of the school and final remarks*, in G. Ciccotti, W.G. Hoover, cit. note 1, pp. 607–610.

## 8.2  Last Opposition

In fact, not withstanding the undeniable contributions of simulation to a better understanding of the properties of condensed matter, one could still hear, even in the period of the Varenna school, echoes of resistance to computational physics by some members of the community of "pure" theoreticians, who, in a more or less direct way, defined as "mere numerical calculations" the activity of their colleagues doing simulation. It is quite possible that, even without explicit mention, Bellemans was thinking of the latest authoritative episode of resistance when he hinted at a charge of "endangering the creativity of the mind". In 1980, the French journal *La Recherche* published a substantial paper by the solid-state physicist Philip Warren Anderson, who had been awarded the Nobel Prize for Physics in 1977, had taught at Princeton, and was director of the theoretical physics group at the prestigious Bell Telephone Laboratories.[4] The paper was entitled *The great illusion of physicists*: Anderson bluntly denounced "the great illusion", in a passionate defence of the identity of a science which, in his own words, was "facing an intellectual challenge which it tries systematically to escape". Following the formulation and the growing success of quantum mechanics and statistical mechanics, Anderson thought that three different directions had taken shape within the physics community. The first consisted in abandoning the field of the structure of matter at the macroscopic scale, where one assumed that, in principle, all fundamental laws had been understood, and concentrating on the study of still poorly explored domains (the infinitely small world of elementary particles, or the infinitely large area of cosmology). The second direction, supported by Anderson, consisted in recognizing that, given the complexity of the concrete structures in which matter is organized, there must be new laws and concepts to be discovered: this was the noble task which the theoretical physics of matter should pursue. The third direction was followed by those who, convinced that there was nothing new to be discovered and that we should simply apply well-known laws, confined their activity to "pure and simple calculation".

This would be the great illusion of physicists: «the followers of this direction accept the idea that all the main theoretical principles have already been found and understood. There is only one thing we can do, namely: calculate, by using a certain number of "fundamental" equations [...] In order to explain a phenomenon, one should simply write down conveniently the set of equations, and there you are! This is what I call the syndrome of the "enchanted machine", since you only need a computer. These physicists imagine that the interpretation of all interesting physical phenomena will turn up from a dream computer».

In short, Anderson's arguments can be reduced to two principles of impossibility. The first is a principle of practical reason: no matter how powerful computers are now and will be in the future, Anderson said, they will never be able to handle all the information contained into a little piece of solid matter, which is made up of a huge number (of the order of ten thousand billion billions) of interacting molecules. However, this is not the real problem. Rather, the point is a second principle of

---

[4]P.W. Anderson, *La grande illusion des physiciens*, La Recherche **107**, 98, 1980.

pure reason, so to speak: in nature "the whole is more than the sum of its parts", as Anderson likes to say. This is a way of saying that macroscopic phenomena are not a mere sum of individual atomic micro-phenomena. When a large number of atoms and molecules are assembled, "emergent phenomena" take place, which cannot generally be foreseen a priori. In sum, within macroscopic systems, what most matters are the collective properties of matter. These properties cannot be foreseen on the basis of the physical and chemical features of individual elements.

As a matter of principle, it would therefore be a "great illusion" if physicists thought they could explain and predict macroscopic phenomena, starting from elementary particles and from the fundamental laws of physics, recognizing computing power as the only limit. This is not true. If we want to understand nature at a macroscopic level, Anderson writes, we need first of all creativity and imagination. These are qualities that computers do not possess.

On closer examination, the target of Anderson's raging polemic was limited to computational quantum chemistry. It is significant that, throughout his paper, the word "simulation" is nowhere to be found. In his invective against the "enchanted machine", Anderson seemed to be completely unaware of the fact that the recourse to computer power in the physics of matter had not been limited—and long since!—to adding a few significant numbers to the calculation of the energy level of an atomic structure. Moreover, when Anderson came to give concrete examples of the alleged inability of "those making calculations" to answer pressing questions, he ended up choosing the wrong cases. This may well have passed unnoticed at the time, but appears conspicuously *ex post*. In order to illustrate his general principle (utterly worthy of support), according to which "the whole is more than the sum of its parts", which would therefore make unrealistic any effort to reconstruct the whole starting from its building blocks, Anderson provided examples of liquid crystals and phase transitions, arguing that "the calculating physics" could not obtain interesting results, "unless appropriate techniques are introduced". As we have see, speaking of liquid crystals and phase transitions, simulation had just achieved, or was about to achieve, very interesting results, thanks to the "introduction of appropriate techniques", in which computing power was blended with the "creativity and imagination" of good theoretical physics.

In condensed matter physics, Anderson said, the macroscopic blocks of matter studied have to be regarded as infinite from the viewpoint of statistical mechanics. With a numerical approach, this difficulty can easily be avoided when one can exploit global symmetry properties, as in a regular crystal, assuming the structure to be regular and periodic. However, «this fine logic falls apart as we try to know whether we are still in a crystal phase, or if some change has occurred. These challenges come up when we study melting, or the relative stability of two crystal phases, or else the physics of liquid crystals. In its liquid phase, the structure is no longer orderly, and there is no point anymore in speaking of global symmetry» . This is true. However, the first papers of Daan Frenkel on liquid crystals were showing in that very period that this limitation on the computational approach was actually being overcome. It is also true, as Anderson himself emphasized, that there had been remarkable theoretical progress in understanding symmetry-breaking in the phase transitions occurring in the interpretation of anti-ferromagnetism, superfluidity, and superconductivity. However, the fact that it was impossible to understand those phenomena starting from models which only took into consideration a few atoms should not hide the fact that Kurt Binder—in those very months—had succeeded in modelling that class of phenomena, and had used simulation to calculate the values of the critical parameters in situations in which "all atoms take part, at the same time and all together, in the phenomenon's mechanism", a feature which, according to Anderson, would foreclose in principle any hope of the "calculating physics" gaining any result.

In sum, Anderson seemed to forget, in his rage against the intellectual inadequacy which would mark the recourse to computers as "enchanted machines", the epistemological novelty introduced when computers were employed in simulation as part of the theory, rather then as mere quick processors of complex calculations. On the other hand, this same novelty had apparently not been grasped by another well-known protagonist, Richard Feynman, who, in a paper with an otherwise promising title, turned his attention to the problem of the use of computers in theoretical physics.[5] Unlike Anderson, Feynman did not want to enter a controversial debate on the intensive use of calculation. Indeed, he seemed to be expecting to learn a lot from the development of computers. However, when he talked of "simulating physics with computers", he was interested in the following question: can we really discover something new about the laws of nature which we do not know as yet, thanks to the computer? His review on the state-of-the-art focussed upon the chances offered by the development of quantum computers, considering that the new laws of physics that had still to be discovered belonged to that domain. In his dense paper on physics, simulation, and computers, Feynman did not even mention simulation in statistical mechanics, either classical or quantum. Moreover, when he introduced his considerations, he wrote that «of course, if we thought that we perfectly know all the laws of physics, we should not pay attention to computers». Saying this, Feynman seemed simply to be ignoring the perspective in which simulation had originated and grown: even if one thought that one already knew all the laws, the explanations of the contents of these laws, and therefore the ability to control nature (the very goal of theoretical physics) requires precisely "paying attention to computers", since they are the tools enabling this wish to come true.

However, thirty years after its birth, molecular simulation had reached such a level of development that the last conceptual objections to its epistemic validity appeared more like a physiological distrust of the older generation towards novelty than solid arguments. In this case, too, we might aptly quote Max Planck, according to whom new ideas in science get firmly established, not so much because of their intrinsic validity, but because a generation used to old ideas disappears and is replaced by a new one which has been brought up with the new ideas. In Varenna, Bellemans addressed himself to this new generation. He merely mentioned in passing the objections which

---

[5]R.P. Feynman, *Simulating physics with computers*, International Journal of Theoretical Physics **21**, 467, 1982.

were being raised against the "calculating physics", since those listening to his talk already knew that they had the right credentials.

Based on what we said above, we can conveniently take the Varenna school as an end point for our story, which aims at a reconstruction of the birth and consolidation (both intellectual and institutional) of molecular simulation. Naturally, when looking in retrospect at the intervening years, we could also consider it as a starting point. Indeed, a very successful annual summer school was set up in the UK in 1988, sponsored by the CCP5 and directed by Mike Allen, with the aim of training a new generation of computational scientists year by year. Shortly after that, in 1990, another celebrated annual summer school was born in Amsterdam, at the initiative of Daan Frenkel and his group. These schools are still alive, in part supported by CECAM. Moreover, starting from the mid-80s, simulation, by then established as a research tool for theoretical physics, set off on the path to exponential growth: until 1986, the number of published papers related to "molecular simulation" was of the order of 500, according to the authoritative source "Web of Science". This number grew to about 14,000 between 1987 and 1996, to reach almost 50,000 between 1997 and 2006, and further doubled in the following decade.

A further sign of this growth is provided by looking at the proceedings of two further schools devoted to molecular simulation, held respectively ten and twenty years after the Varenna school, in the same spirit of assessing the state of the discipline, and discussing the evolution of techniques and current and future applications. For organizational reasons, the number of participants, both lecturers and students, never exceeded the first Varenna experience, neither in Como in 1995, nor in Erice in 2005. On the other hand, the number of pages of the volumes of the proceedings was expanding: in the case of Como, all contributions could still be included in a single 1000-page volume, whereas, ten years later, two separate volumes had to be published, for a total of about 1500 pages, devoted to explaining fundamental techniques and applications, respectively.[6] Moreover, while in Como it was still possible to cover the whole range of techniques and potential developments in one meeting, ten years later the exponential growth of this sector imposed drastic cuts on the issues that could reasonably be treated, including both theory and applications. With the Erice school, it had become clear that the field had grown so much that these kinds of all-encompassing events could not go on.

However, this is recent history. We shall stop at the point at which the novelty was clearly established. As a conclusion to our story, we can go back to a few of our initial points, in order to better characterize certain distinctive traits of this novelty.

---

[6]K. Binder, G. Ciccotti (eds.), *Monte Carlo and molecular dynamics of condensed matter systems: Euroconference on Computer Simulation in Condensed Matter Physics and Chemistry: Como, 3–28 July 1995*, SIF, Bologna 1996; M. Ferrario, G. Ciccotti, and K. Binder (eds.), *"Computer Simulations in Condensed Matter: From Materials to Chemical Biology"*, LNP Vol. 1&2, Springer, Berlin, 2006.

## 8.3   A Scientific Revolution?

Novelties are created all the time during the growth of any scientific discipline. Quite simply, without novelties our knowledge would never increase. However, the very word "novelty" is too vague. We may consider both the discovery of a new sub-species of plant parasite and the discovery of the DNA replication mechanism as novelties, but it is obvious that these two novelties imply quite different shifts in the structure of knowledge. It is common practice to characterize "great" novelties, namely those that mark a significant turning point in a particular discipline, labelling them as "scientific revolutions". This allows us to draw a clear distinction between parasite and DNA, but does not necessarily solve the problem of setting clear boundaries between general advancements and revolutionary innovations. The point is that even the term "revolution" (from now on, for the sake of simplicity, we shall not specify that we are talking about "scientific revolutions") has not been given a well defined meaning. Scientists have often spoken about "revolutions" in their own or other people's disciplines. Indeed, they have used this definition more and more, mainly since the beginning of the twentieth century, in an increasingly nonchalant fashion. Supporters of the electromagnetic worldview who opposed relativity at the beginning of the twentieth century explicitly called themselves "revolutionaries", while relativity subsequently became, in its turn, the prototype of a revolutionary theory. Later on, during its development, the revolution of quantum theory actually produced a series of revolutions (from the old quantum theory to quantum mechanics, to the quantum theory of fields), down to the revolution of transistors in solid state physics and the "November revolution" in high-energy physics.

To bring order to this chaos and use the term "revolution" in a stricter and less casual way, we can refer to the well-known work of Thomas Kuhn, who gave the scientific revolution a definite epistemological identity.[7] This is not the place to delve into an exegesis of Kuhn's thought, nor to describe the extensive debate which, over the last half century, has troubled—and still troubles—the milieu of the philosophy of science. It is enough for us to sum up what Kuhn considers to be the distinctive traits of a revolution: the affirmation of a new way of looking at the world of nature, which cannot be matched with the preceding one, while, at the same time, new tools (experimental equipment, formal language, theoretical constructs) are strengthened in order to take a new look at the world. A scientific revolution takes place through the establishment of new paradigms replacing old ones and creating a frame of mind which views the world according to a brand-new perspective. Thus, a new scientific practice is also produced and shared.

---

[7]T. Kuhn, *The Structure of Scientific Revolution*, Chicago University Press, Chicago 1962.

Not by chance, the philosophical-epistemological concepts underlying *The structure of scientific revolutions* appeared a few years after the publication of Kuhn's fine work on the history of the Copernican revolution.[8] There is little doubt that, in this particular case, our view of the world changed dramatically, and a model was proposed which was completely different from the preceding one. Kuhn's categories of incommensurability and paradigm shift seem to be perfectly adapted to characterizing this transition. And so much more so if we extend our perspective to the revolutionary (and the word is appropriate here) change in life, culture, and society at large right through the rest of century. As far as science is concerned, we can sum it up, well aware of the schematization involved here, by mentioning the names of Galileo and Newton. Thanks to their work, the new worldview arising in the sixteenth century created the intellectual tools needed to pass from contemplation of nature to active intervention, validating new instruments and adopting new paradigms. As far as material instruments are concerned, Galileo's novelty did not consist so much in his using an eyepiece to look at the sky, as in deciding that this use was a legitimate cognitive procedure; this is what really made his recourse to a readily available instrument, which had already been used for other purposes, a genuine revolutionary step. The same can be said for his methodological statement that to know nature you have to learn to read it through the characters in which it is written, and to speak the mathematical language which reveals its hidden structure: «Philosophy is written in this grand book, which stands continually open before our eyes (I say the Universe), but can not be understood without first learning to comprehend the language and know the characters as it is written. It is written in mathematical language, and its characters are triangles, circles, and other geometric figures, without which it is impossible to humanly understand a word; without these one is wandering in a dark labyrinth».[9]

Seventy years later, Isaac Newton said it again: «A vulgar Mechanik can practice what he has been taught or seen done, but if he is in an error he knows not how to find it out and correct it, and if you put him out of his road, he is at a stand; whereas he that is able to reason nimbly and judiciously about figure, force and motion, is never at rest till he gets over every rub».[10]

---

[8]T. Kuhn, *The Copernican Revolution. Planetary Astronomy in the Development of Western Thought*, Harvard University Press, Cambridge, Mass. 1957.

[9]G. Galilei, *Il Saggiatore*, Roma, Giacomo Mascardi 1623, p. 232. Translated by T. Salusbury, 1661.

[10]I. Newton to N. Hawes, 25 May 1694, in J. Edleston, ed., *Correspondence of Sir Isaac Newton and Professor Cotes*, London, J.W. Parker 1850, p. 284.

> *A Vulgar Mechanick can practice what he has been taught or seen done, but if he is in an error he knows not how to find it out and correct it, and if you put him out of his road, he is at a stand; Whereas he that is able to reason nimbly and judiciously about figure, force and motion, is never at rest till he gets over every rub.*
>
> Isaac Newton to Nathaniel Hawes
> 25 May 1694

Isaac Newton on the difference between empirical know-how and scientific knowledge (Cambridge Univ. Press)

We can therefore say that the scientific revolution of the seventeenth century—led by Galileo and Newton—constitutes a real revolution, since it has radically changed the way we look at and get to know the Universe, as Alexandre Koyré recalls, passing "from a closed world to an infinite Universe", "from the world of approximation to the Universe of precision".[11] «These novelties about old truths, new worlds, new stars, new systems, new nations, etc., are the start of a new century», wrote Tommaso Campanella, fully conscious of the dramatic transition brought about at the beginning of the seventeenth century by Galileo's astronomical observations.[12] In particular, this was an authentic scientific revolution—an epistemological revolution—since it changed not only the picture of the world (i.e., Copernicus' revolution), but also the means to create and validate that picture, establishing the idea that the fundamental laws of nature should not be inferred from "observations" (i.e., passive recording of natural phenomena), but rather from "experiments" (i.e., active questioning of nature, led by theoretical abstractions), all expressed in the language of mathematics, the only language which can unambiguously characterize their meaning. If the world is written in the language of mathematics, the laws we discover can be used to predict, through calculation, the evolution of specific natural phenomena; in principle, of all natural phenomena. This was a lucky and revolutionary intuition. Indeed, the same scientists who started the process were well aware that it was a clean break with the essentially speculative approach which had till then held sway in natural philosophy. In fact it was. Isaac Newton brought this revolution to completion, going beyond Galileo and Bacon, and introducing the most suitable instrument for the mathematization of physics, namely infinitesimal calculus.

---

[11] A. Koyré, *From the Closed World to the Infinite Universe*, Baltimore, Johns Hopkins Press 1957.

[12] T. Campanella to G. Galilei, August 5, 1632. Florence, Biblioteca Nazionale Centrale, Galileiani 92, ff. 224–225 (author's translation).

The scientifically mature acquisition of the transformation we are describing fully represents the shift in consciousness which emerged at the end of the Middle Ages. Eugenio Garin described this moment in an elegant way: «The medieval perspective had really exhausted all possible implications of the classical outlook; it had reached its saturation point. Yet, at that time in the complex 14th century, between its last word and the first word of the new way of thinking, there was the same sort of divergence that characterizes, to use a common way of putting things in those days, the state of rest and the first instant of motion, the last moment of illness and the first moment of health, the last breath and the first moment of death: there was a jump. This was the passage from the vision of the being enclosed in its own reality to the man as poet, i.e. creator; to the man who does not merely contemplate a given natural order, to achieve an eternal essence, but who can now grasp a world of infinite possibilities, who *is himself* an infinity of possibilities. The world, far from being crystallized in a series of immutable forms, becomes moldable into always novel ways, and there is now no necessity that can not be broken, there is now no form that can not be transformed, and freedom of man now means a being whose face would never be defined».[13]

Now, we may say that nothing of the kind occurred in the latest development of physics; in any case, nothing which would count as a similarly radical epistemic leap away from the previous conceptual framework. Through all the changes and the new ideas which have marked the story of this discipline, from the seventeenth century onwards, physicists basically kept working within the conceptual framework of Galileo and Newton, grounding their research upon the derivation of fundamental laws from accurate observations and experiments, and expressing the laws thus derived in mathematical terms.

This is also true of the "new physics" of the twentieth century, no matter how often it has represented itself as "revolutionary". Even though the conquests realized by relativity and quantum mechanics were relevant and ground-breaking, they never deeply affected our way of knowing nature. Indeed, they did not change our epistemology. Nor have they even been accompanied by a transformation of society, or at least, by a transformation of social consciousness. To be sure, their construction involved important paradigm shifts; these were conceptually traumatic novelties, in comparison with the prevailing assumptions in the traditional mechanistic theory (i.e., classical mechanics). However, they did not constitute a revolution in the sense

---

[13] Our translation from: E. Garin, *Medioevo e Rinascimento: Studi e Ricerche*, Laterza, Bari 1954, pp. 37–38: «L'analisi medievale aveva, a tal punto, esaurito veramente tutte le possibilità delle impostazioni classiche; era veramente giunta al limite. Ma fra l'ultima sua parola e la prima delle nuove correnti sbocciate, non dimentichiamolo, in quel medesimo e così complesso Trecento, v'è la stessa ossessionante divergenza che corre – per prendere un tema allora d'uso – fra l'ultimo istante della quiete di un corpo e il primo del suo movimento, fra l'ultimo momento della malattia e il primo della sanità, fra l'ultimo respiro di vita e il primo punto di morte: c'è un salto. C'è un passaggio dalla visione dell'essere conchiuso nella sua realtà all'uomo poeta, che vuol dire creatore. All'uomo che non ha da contemplare un ordine dato, da attuare un'essenza eterna, ma che ha dinanzi infinite possibilità; che è infinite possibilità. Il mondo, lungi dall'essere fisso entro forme cristallizzate, è plasmabile in guise sempre nuove, e non c'è necessità che non s'incrini, non forma che non si trasformi; e libertà d'uomo indica un essere il cui volto non è mai definito».

which we have tried to explain, which seems to us to be the real meaning of revolutionary change, as suggested by Kuhn. And incidentally, Einstein himself, talking about his theories of special and general relativity, denied that they constituted a "revolution". Rather, he thought they were just part of a process of generalization of Newton's theory of mechanics. Of course, we now know that, when we look at the Universe, we should modify our ideas about the structure of space-time, and use a mathematical language which is more sophisticated than the one used to describe ordinary mechanics. And it is now clear that we can no longer think of an atom as a miniature planetary system, and that we have to refer to more abstract concepts and better suited formal structures to model the microscopic world. However, the nature of the knowledge process has not changed: infer from observation (after removing "impediments") all general and reproducible behaviour, translate it into mathematical laws and equations, which we assume to describe general categories of phenomena, and submit it to calculation in order to test its reliability and predict and control observable behaviour. At least, this is what we do whenever we can.

That's where the novelty of simulation comes in. Before modern computers appeared, the task of forecasting and controlling natural phenomena, potentially implied by the equations that were supposed to describe them, actually remained confined to a very limited number of cases. We know that the reason for this limited capacity to fully exploit the potentials offered, in principle, by the mathematical representation of natural laws is essentially our very poor computing power. This limitation was somehow accepted by the long-prevailing attitude in theoretical physics which privileged the discovery of laws over the explanation of their contents. Understanding the world meant essentially discovering the fundamental laws ruling its behaviour; once these laws had been established, at a given scale, the task of physics had been fulfilled and we could move on to discover other, as yet unknown laws at different scales (passing, for example, from atoms to nuclei, or from planets and the Solar System to the large-scale Universe), and leaving to others the problem of trying to actually apply the well-known laws to complex situations. In so doing, physics was placing itself at the top of the epistemic hierarchy of sciences, while at the same time failing to fulfil a crucial part of the plan according to which it had been conceived, namely *understanding, reconstructing,* and *controlling* the universe of phenomena; it reserved for itself the "noble" part of understanding, while delegating to others the vile task of reconstructing and controlling.

With the invention and rapid development of electronic computers, the situation changed radically. Just as the eyepiece expanded the observational capacity of our human senses and gave us the opportunity to see new worlds and build new theoretical views of nature, computers, thanks to their continuously improving performances, provided a processing power which allows scientists to "see" in ever greater depth what is implicitly contained in their equations, and complete the ambition of physics of explicitly predicting, and hence controlling, the behaviour of systems that no one had been able to handle until then. Fundamental laws can at last be used to effectively simulate the real world in all its complexity; one is no longer confined to show that the behaviour of the real material world is consistent with general laws, for it can also be demonstrated that this behaviour can be foreseen on the basis of those very laws, whose implicit contents can now be fully developed.

One can say, more specifically, that simulation is an epistemic novelty, not so much because it represents a "third way" between "sensed experiences" and "certain demonstrations", between experiment and theory, but because it constitutes a novelty within physical theory. Notwithstanding the caution of the founding fathers, who took care to talk about "computer experiments", there is nothing more "experimental" in simulation than the use of a computer, which—obviously—is a material object rather than an intellectual concept; it is the equivalent of the paper and pencil used by a "classical" theoretical physicist. However, what we observe with a computer, what a computer allows us to study, is not a piece of matter, as it would be in an experiment; a computer calculates and provides information about the behaviour of a theoretical model it has been presented with. Simulation is physical theory. The novelty consists in the fact that it allows physical theory to explore such fields, like chemical and biological complexity, for instance, which used to be off-limits. In other words, simulation constitutes a novelty within theoretical physics, because it is a new exercise, which has been made possible by computing power; which starts from the well established laws of physics, translates them into proper algorithms, and proceeds to calculate the behaviour of complex systems.

Indeed, the goal of fundamental simulation consists in building up realistic fundamental models of complex physical systems, and simulating their behaviour according to the fundamental laws of physics. In other words, fundamental simulation on a computer allows scientists to model and predict the observable behaviour of real physical systems, starting from first principles, and using as sole input the fundamental laws of physics. Thanks to the development of computers and powerful, effective algorithms, this path leads us directly from the fundamental laws of physics to observable phenomena. This is a straight path, which can be taken up when studying a wider and wider range of physical, chemical, and even biological systems. Thanks to fundamental simulation, we can now predict the structure and properties of systems of unimaginable complexity. This is true from astrophysics and elementary particles to biochemistry and biology.

In particular, molecular simulation is an authentic novelty, inasmuch as it has redefined the traditional method of practising scientific research, providing a way to make realistic predictions about the behaviour of complex molecular systems (in nature and number), which were previously simply off-limits. As a consequence, molecular simulation has reshaped the whole structure of the classification of scientific fields, creating new links among different disciplines, and within those disciplines that were traditionally considered as "not exact", such as medical science or pharmaceutics.

All this is something new. Following Kuhn's categories, it is probably correct to characterize the novelty introduced by simulation as a significant change of paradigm; one starts looking at phenomena in a different way, because unprecedented procedures can be put into practice, and this fixes new rules and defines a new "normal science". At the same time, all this is still taking place within the epistemological framework of Galileo and Newton.

It is true that scientific development is still widely perceived as a "revolutionary" process. However, this is due not so much to a real upheaval of the cognitive processes, as to the evident speed of the accompanying changes, whereby our current world (i.e., the world of knowledge and, in general, the world in which we are living, which science is continuously transforming) appears to us to be radically different from yesterday's world, and our recent past seems to us much farther away than theirs would have seemed to earlier generations. This is the subjective and widespread perception of the effects of exponential growth, which is so fast that it is almost impossible to comprehend, and was actually triggered by the major transformation which kick-started modern science. Change occurs faster and faster, without discontinuity, and there is a growing rate of unprecedented news, so it feels like as though we are facing a real revolution with respect to the pre-existing state of affairs. The more and more often used metaphor of the "scientific revolution" can likely be attributed to this widely felt perception.[14]

Over and above this perception, we may say that there is no epistemological break with the cognitive process established by the scientific revolution which began the process of modern science. On the contrary, one might say that simulation constitutes the culmination of that project, which has finally made true our underlying dream of controlling the world.[15] Therefore, the appearance and diffusion of simulation do not represent yet another revolution. Indeed, taking a closer look and sticking to the meaning which we have tried to define, there has been only one true revolution, and we are still moving in its wake. Instead, what we are witnessing is, in order of time, just the latest act of that revolution, whose potential still remains largely unexplored. It is difficult to foresee where this new predicting power will lead us. It is an on-going process, which certainly has some surprises in store for us.

---

[14] About the perception of change, and the mythology of scientific revolutions, see M. Daumas, *Le cheval de César, ou le mythe des révolutions techniques,* Éditions des Archives contemporaines, Paris 1991.

[15] D. Macuglia, B. Roux, G. Ciccotti, *Sense Experiences and "Necessary Simulations": Four Centuries of Scientific Change from Galileo to Fundamental Computer Simulations*, KNOW: A Journal on the Formation of Knowledge, **4:1**, 63, 2020.

# Bibliography

ADAMS D. J., *Grand Canonical Ensemble Monte Carlo for a Lennard-Jones Fluid*, in "Molecular Physics", 29, 1, 1975, pp. 307-11.

ALDER B. J., *Numerical Experiment in Statistical Mechanics,* in "Computer Physics Communication", 3, Supplement 1, 1972, pp. 86-91.

ID., *Computer Dynamics,* in "Annual Review of Physical Chemistry", 24, 1973, pp. 325-37.

ID, *Concluding Remarks: The Long-Time Tails Story,* in M. Mareschal, B. L. Holian (eds.), *Microscopic Simulations of Complex Hydrodynamic Phenomena*, Plenum Press, New York 1992, pp. 425-30.

ID., *In Memoriam: Thomas E. Wainwright. September 22,1927-November 27,2007,* in "Progress of Theoretical Physics", Supplement 178, 2009, pp. 1-4.

ALDER B. J., CEPERLEY D. M., POLLOCK E. L., *Computer Simulation of Phase Transitions in Classical and Quantum Systems,* in "International Journal of Quantum Chemistry", 22, S16, 1982, pp. 49-61.

ALDER B. J., FRANKEL S., LEWINSON V., *Radial Distribution Function Calculated by the Monte Carlo Method for a Hard Sphere Fluid,* in "The Journal of Chemical Physics", 23, 3, 1955, pp. 417-9.

ALDER B. J., GASS D. M., WAINWRIGHT T. E., *Studies in Molecular Dynamics. viii. The Transport Coefficients for a Hard-Sphere Fluid,* in "The Journal of Chemical Physics", 53, 10, 1970, pp. 3813-26.

ALDER B. J, WAINWRIGHT T. E., *Phase Transition for a Hard Sphere System,* in "The Journal of Chemical Physics", 27, 5, 1957, pp. 1208-9.

ID, *Molecular Dynamics by Electronic Computers,* in I. Prigogine (ed.), *Proceedings of the International Symposium on Transport Processes in Statistical Mechanics*, Proceedings of the Symposium (Brussels, August 27-31, 1956), Interscience Publishers, London 1958, pp. 97-131.

ID., *Molecular Motion,* in "Scientific American", 201, 4, 1959, pp. 113-26.

ID., *Studies in Molecular Dynamics. i. General Method,* in "The Journal of Chemical Physics", 31, 2, 1959, pp. 459-66.

ID., *Investigation of the Many-Body Problem by Electronic Computers,* in J. K. Percus (ed.), *The Many-Body Problem*, Proceedings of the Symposium at Stevens Institute of Technology (Hoboken, nj, January 28-29, 1957), Interscience Publishers, New York-London 1963, pp. 511-22.

ID., *Velocity Autocorrelations for Hard Spheres,* in "Physical Review Letters", 18, 23, 1967, pp. 988-90.

ALDER B. J. et al., *A Molecular Dynamics Study of the Intensity and Band Shape of Depolarized Light Scattered from Rare-Gas Crystals,* in "Physica B+C", 83, 3, 1976, pp. 249-58.

ANDERSEN H. C, *Molecular Dynamics Simulations at Constant Pressure and/or Temperature,* in "The Journal of Chemical Physics", 72, 4, 1980, pp. 2384-93.

© Springer Nature Switzerland AG 2020

G. Battimelli et al., *Computer Meets Theoretical Physics*, The Frontiers Collection, https://doi.org/10.1007/978-3-030-39399-1

ANDERSON J. B., *Statistical Theories of Chemical Reactions. Distributions in the Transition Region*, in "The Journal of Chemical Physics", *58, 10, 1973, pp.* 4684-92.

ANDERSON P. W., *La grande illusion des physiciens*, in "La Recherche", 107, 2, 1980, pp. 98-102.

ASHURST W. T., HOOVER W. G., *Argon Shear Viscosity via a Lennard-Jones Potential with Equilibrium and Nonequilibrium Molecular Dynamics*, in "Physical Review Letters", 31, 4, 1973, pp. 206-8.

BARKER J., *A Quantum-Statistical Monte Carlo Method: Path Integrals with Boundary Conditions*, in "Journal of Chemical Physics", 70, 6, 1979, pp. 2914-8.

BARKER J. A., KLEIN M. L., *Monte Carlo Calculations for Solid and Liquid Argon*, in "Physical Review B", 7, 10, 1973, pp. 4707-12.

BARKER J. A., WATTS R. O., *Structure of Water: A Monte Carlo Calculation*, in "Chemical Physics Letters", 3, 3, 1969, pp. 144-5.

BATTIMELLI G., *An European Research Centre in Orsay: CECAM 1969-1996,*, CECAM, Lausanne 2019.

BATTIMELLI G., CICCOTTI G., *Berni Alder and the Pioneering Times of Molecular Simulation*, in "European Physical Journal H", 43, 303, 2018, pp. 303-35.

BATTIMELLI G., FRENKEL D., *Berni Alder Interview* (June 18, 1990), Niels Bohr Library, American Institute of Physics 1990.

BAUMGÄRTNER A., BINDER K., *Monte Carlo Studies on the Freely Jointed Polymer Chain with Excluded Volume Interaction*, in "The Journal of Chemical Physics", 71, 6, 1979, pp. 2541-5.

BELLEMANS A., *Synopsis of the School and Final Remarks*, in G. Ciccotti, W. G. Hoover, *Molecular-Dynamics Simulation of Statistical-Mechanical Systems*, Proceedings of the "Enrico Fermi" International Summer School of Physics, Course XCVII (Varenna, July 23-August 2, 1985), North-Holland, Amsterdam 1986, pp. 607-10.

BENNETT C. H., *Mass Tensor Molecular Dynamics*, in "Journal of Computational Physics", 19, 3, 1975, pp. 267-79.

ID., *Efficient Estimation of Free Energy Differences from Monte Carlo Data*, in "Journal of Computational Physics", 22, 2, 1976, pp. 245-68.

BENNETT C. H., ALDER B. J., *Studies in Molecular Dynamics. IX. Vacancies in Hard Sphere Crystals*, in "The Journal of Chemical Physics", 54, 11, 1971, pp. 4796-808.

BERENDSEN H. J. C., *Introduction*, in *Models for Protein Dynamics*, CECAM Workshop (Orsay, May 24-July 17, 1976), CECAM, Orsay 1976.

ID., *The Development of Molecular Dynamics at CECAM*, in J. P. Hansen, G. Ciccotti, H. J. C. Berendsen, *In Memoriam: Aneesur Rahman 1927-1987*, CECAM Meeting (Orsay, September 11, 1987), CECAM, Orsay 1987, pp. 9-12.

BINDER K., *A Monte Carlo Method for the Calculation of the Magnetization of the Classical Heisenberg Model*, in "Physics Letters A", 30, 5, 1969, pp. 273-4.

ID., *Monte Carlo Study of Thin Magnetic Ising Films*, in "Thin Solid Films", 20, 2, 1974, pp. 367-81.

ID., *Finite Size Analysis of Ising Model Block Distribution Functions*, in "Zeitschrift für Physik B", 43, 1981, 119-40.

ID., *From the Lake of Como to Surface Critical Phenomena*, in J. Brujic, A. Grosberg (eds.), *Memories of Pierre Hohenberg*, New York University Press, New York 2018, pp. 30-4.

ID.,, (ed.), *Monte Carlo Methods in Statistical Physics*, Springer-Verlag, Berlin 1979.

BINDER K., CICCOTTI G. (eds.), *Monte Carlo and Molecular Dynamics of Condensed Matter Systems: Euroconference on Computer Simulation in Condensed Matter Physics and Chemistry*, Proceedings of Symposium (Como, July 3-28, 1995), Sif, Bologna 1996.

BINDER K., HOHENBERG P. C., *Phase Transitions and Static Spin Correlations in Ising Models with Free Surfaces*, in "Physical Review B", 6, 9, 1972, pp. 3461-87.

ID., *Surface Effects on Magnetic Phase Transitions*, in "Physical Review B", 9, 5, 1974, pp. 2194-214.

BINDER K., LANDAU D. P., *Critical Properties of the Two-Dimensional Anisotropic Heisenberg Model*, in "Physical Review B", 13, 3, 1976, pp. 1140-55.

BINDER K., MULLER- KRUMBHAAR H., *Monte Carlo Calculation of the Scaling Equation of State for the Classical Heisenberg Ferromagnet*, in "Physical Review B", 7, 7, 1973, pp. 3297-306.

BINDER K., RAUCH H., *Calculation of Spin-Correlation Functions in a Ferromagnet with a Monte Carlo Method*, in "Physics Letters A", 27, 4, 1968, pp. 247-8.

ID., *Numerische Berechnung von Spin-Korrelationsfunktionen und Magnetisierung-skurven von Ferromagnetica*, in "Zeitschrift für Physik", 219, 3, 1969, pp. 201-15.

BINDER K., SCHRODER K., *Monte Carlo Study of a Two-Dimensional Ising "Spin Glass"*, in "Solid State Communications", 18, 9-10, 1976, pp. 1361-4.

ID., *Phase Transitions of a Nearest-Neighbor Ising "Spin Glass"*, in "Physical Review B", 14, 5, 1976, pp. 2142-52.

BINDER K., STAUFFER D., *Monte Carlo Study of the Surface Area of Liquid Droplets*, in "Journal of Statistical Physics", 6, 1, 1972, pp. 49-59.

ID., *Statistical Theory of Nucleation, Condensation and Coagulation*, in "Advances of Physics", 25, 4, 1976, pp. 343-96.

BINDER K., STAUFFER D ., MULLER- KRUMBHAAR H., *Calculation of Dynamic Critical Properties from a Cluster Reaction Theory*, in "Physical Review B", 10, 9, 1974, pp. 3853-7.

BINDER K. et al., *Computer Experiments on Phase Separation in Binary Alloys*, in "Advances in Colloid and Interface", 10, 1, 1979, pp. 173-214.

BRUSH S., SAHLIN H. L., TELLER E., *Monte Carlo Study of a One-Component Plasma. I*, in "The Journal of Chemical Physics", 45, 6, 1966, pp. 2102-18.

CANGUILHEM G. (ed.), *La mathématisation des doctrines informes*, Hermann, Paris 1972.

CAR R., PARRINELLO M., *Unified Approach for Molecular Dynamics and Density-Functional Theory*, in "Physical Review Letters", 55, 22, 1985, pp. 2471-4.

CARTER E. A. et al., *Constrained Reaction Coordinate Dynamics for the Simulation of Rare Events*, in "Chemical Physics Letters", 156, 5, 1989, pp. 472-7.

CEPERLEY D. M., *Ground State of the Fermion One-Component Plasma: A Monte Carlo Study in Two and Three Dimensions*, in "Physical Review B", 18, 7, 1978, pp. 3126-38.

ID., *Fermion Nodes*, in "Journal of Statistical Physics", 63, 5-6, 1991, pp. 1237-67.

ID., *Path-Integral Calculations of Normal Liquid $^3He$*, in "Physical Review Letters", 69, 2, 1992, pp. 331-4.

ID., *Path Integrals in the Theory of Condensed Helium*, in "Review of Modern Physics", 67, 2, 1995, pp. 279-355.

CEPERLEY D. M., ALDER B. J., *Ground State of Electron Gas by a Stochastic Method*, in "Physical Review Letters", 45, 7, 1980, pp. 566-9.

ID., *The Calculation of the Properties of Metallic Hydrogen using Monte Carlo*, in "Physica B+C", 108, 1981, pp. 875-6;

ID, *Ground State of Solid Hydrogen at High Pressure*, in "Physical Review B", 36, 4, 1987, pp. 2092-106.

CEPERLEY D. M., JACUCCI G, *Calculation of Exchange Frequencies in $bcc^3He$ with the Path Integral Monte Carlo Method*, in "Physical Review Letters", 58, 16, 1987, pp. 1648-51.

CEPERLEY D. M., KALOS M. H., *Quantum Many-Body Problems*, in K. Binder (ed.), *Monte Carlo Methods in Statistical Physics*, Springer, Berlin 1979, pp. 145-94.

CEPERLEY D. M., POLLOCK E. L., *Path-Integral Computation of the Low-Temperature Properties of Liquid $^4He$*, in "Physical Review Letters", 56, 4, 1986, pp. 351-4.

CHANDLER D., *Statistical Mechanics of Isomerization Dynamics in Liquids and the Transition State Approximation*, in "The Journal of Chemical Physics", 68, 6, 1978, pp. 2959-70.

CHANDLER D., WOLYNES P. G., *Exploiting the Isomorphism between Quantum Theory and Classical Statistical Mechanics of Polyatomic Fluids*, in "The Journal of Chemical Physics", 74, 7, 1981, pp. 4078-95.

CICCOTTI G., *The Computer, Machine of the Dreams of Theoretical Physics*, in "SIMU Newsletter", 1, 2000, pp. 5-11.

CICCOTTI G., FERRARIO M., *Non-Equilibrium by Molecular Dynamics: a Dynamical Approach*, in "Molecular Simulation", 42, 16, 2016, pp. 1385-400.

CICCOTTI G., HOOVER W. G. (eds.), *Molecular-Dynamics Simulation of Statistical-Mechanical Systems*, Proceedings of the "Enrico Fermi" International Summer School of Physics, Course XCVII (Varenna, July 23-August 2, 1985), North-Holland, Amsterdam 1986.

CICCOTTI G., JACUCCI G., *Direct Computation of Dynamical Response by Molecular Dynamics: The Mobility of a Charged Lennard-Jones Particle*, in "Physical Review Letters", 35, 12, 1975, pp. 789-92.

CICCOTTI G. et al., *Constrained Molecular Dynamics and the Mean Potential for an Ion Pair in a Polar Solvent*, in "Journal of Chemical Physics", 129, 2, 1973, pp. 241-51.

CICCOTTI G. et al., *Molecular Dynamics Simulation of Ion Association Reactions in a Polar Solvent*, in "Journal de Chimie Physique", 85, 1988, pp. 925-9.

DAUMAS M., *Le Cheval de Cesar, ou le Mythe des Revolutions Techniques*, Editions des Archives contemporaines, Paris 1991.

EDLESTON J. (ed.), *Correspondence of Sir Isaac Newton and Professor Cotes*, J. W. Parker, London 1850.

FERRARIO M., CICCOTTI G., BINDER K. (eds.), *Computer Simulations in Condensed Matter: From Materials to Chemical Biology*, 2 voll., Springer, Berlin 2006.

FEYNMAN R. P., *The λ-Transition in Liquid Helium*, in "Physical Review", 90, 6, 1953, pp. 1116-7.

ID., *Atomic Theory of Liquid Helium Near Absolute Zero*, in "Physical Review", 91, 6, 1953, pp. 1301-8.

ID., *Atomic Theory of the λ Transition in Helium*, in "Physical Review", 91, 6, 1953, pp. 1291-301.

ID., *Simulating Physics with Computers*, in "International Journal of Theoretical Physics", 21, 6-7, 1982, pp. 467-88.

FISHER M. E., *The Theory of Equilibrium Critical Phenomena*, in "Reports on Progress in Physics", 30, 2, 1967, pp. 615-730.

FLORY P. J., *Thermodynamics of High Polymer Solutions*, in "The Journal of Chemical Physics", 9, 8, 1941, p. 660.

FRENKEL D, s.t., in *Recollections of CECAM for Carl*, CECAM, Paris 1990, p. 30.

FRENKEL D., LADD A. J. C., *New Monte Carlo Method to Compute the Free Energy of Arbitrary Solids: Application to the FCC and HCP Phases of Hard Spheres*, in "The Journal of Chemical Physics", 81, 1, 1984, pp. 3188-93.

FRENKEL D., MAGUIRE J. F., *Molecular Dynamics Study of the Dynamical Properties of an Assembly of Infinitely Thin Hard Rods*, in "Molecular Physics", 49, 3, 1983, *pp. 503-41*.

FRENKEL D., MCTAGUE J. P., *Evidence for an Orientationally Ordered Two Dimensional Fluid Phase from Molecular Dynamics Calculations*, in "Physical Review Letters", 42, 24, 1979, pp. 1632-5.

FRENKEL D., MULDER B. M., MCTAGUE J. P., *Phase Diagram of a System of Hard Ellipsoids*, in "Physical Review Letters", 52, 4, 1984, pp. 287-90.

ID., *Phase Diagram of Hard Ellipsoids of Revolution*, in "Molecular Crystals and Liquid Crystals", 123, 1985, pp. 119-28.

GALILEI G., *Il Saggiatore*, Giacomo Mascardi, Roma 1623.

ID., *Dialogo sopra i due massimi sistemi del mondo Tolemaico e Copernicano*, Landini, Florence 1632. Translated by Stillman Drake, *Dialogue Concerning the Two World Systems—Ptolemaic and Copernican*, University of California Press 1967.

GARIN E., *Medioevo e Rinascimento: Studi e Ricerche*, Laterza, Bari 1954.

GEIGER A., *The Simulation of Water and Aqueous Solutions by Aneesur Rahman*, in J. P. Hansen, G. Ciccotti, H. J. C. Berendsen, *In Memoriam: Aneesur Rahman 1927-1987*, CECAM Meeting (Orsay, September 11, 1987), CECAM, Orsay 1987, p. 21.

GIBSON J. B. et al., *Dynamics of Radiation Damage*, in "Physical Review", 120, 4, 1960, pp. 1229-53.

GOLDSTINE H., *The Computer from Pascal to von Neumann*, Princeton University Press, Princeton (nj) 1993 (I$^{st}$ ed. 1972).

GOSLING E. M., MCDONALD I. R., SINGER K., *On the Calculation by Molecular Dynamics of the Shear Viscosity of a Simple Fluid*, in "Molecular Physics", 26, 6, 1973, pp. 1475-84.

GREEN S. M. (ed.), *Critical Phenomena*, Proceedings of the "Enrico Fermi" International School of Physics, Course LI (Varenna, July 27-August 8, 1970), North-Holland, Amsterdam 1970.

GUBERNATIS J. E., *Marshall Rosenbluth and the Metropolis Algorithm*, in "Physics of Plasmas", 12, 057303-1, 2005.

HANSEN J. P., KLEIN M. L., *Computer «Experiments» on Solid Rare Gases: the Dynamical Structure Factor S ( $\vec{Q}$, $\Omega$ )*, in "Journal de Physique Lettres", 35, 3, 1974, pp. 29-31.

ID., *Dynamical Structure Factor S ( $\vec{Q}$, $\Omega$ ) of Rare-Gas Solids*, in "Physical Review B", 13, 2, 1976, pp. 878-87.

HANSEN J., VERLET L., *Phase Transitions of the Lennard-Jones System*, in "Physical Review", 184, I, 1969, pp. 151-61.

HARP G. D., BERNE B. J., *Time-Correlation Functions, Memory Functions, and Molecular Dynamics*, in "Physical Review A", 2, 3, 1970, pp. 975-96.

HOOVER W.G., *Canonical Dynamics: Equilibrium Phase-Space Distributions*, in "Physical Review A", 31, 3, 1985, pp. 1695-7.

ID., *NOSÉ SHUICHI, 17 June 1951-17 August 200s, In Memoriam*, in "Butsuri", 60, 10, 2005, p. 819 (http://www.williamhoover.info/nose.pdf)

ID., *From Ann Arbor to Sheffield: Around the World in 80 Years. 1. Yokohama to Ruby Valley*, in "Computational Methods in Science and Technology", 23, 2017, pp. 133-41.

HOOVER W. G., ASHURST W. T., *Nonequilibrium Molecular Dynamics*, in "Theoretical Chemistry: Advances and Perspectives", I, 1975, pp. 1-51.

HOOVER W. G., REE F. H, *Use of Computer Experiments to Locate the Melting Transition and Calculate the Entropy in the Solid Phase*, in "The Journal of Chemical Physics", 47, 12, 1967, pp. 4873-8.

ID., *Melting Transition and Communal Entropy for Hard Spheres*, in *"The Journal* of Chemical Physics", 49, 8, 1968, pp. 3609-17.

HUANG K., *Statistical Mechanics*, John Wiley & Sons, New York 1963.

HUGGINS M. L., *Solutions of Long Chain Compounds*, in "The Journal of Chemical Physics", 9, 5, 1941, p. 440.

JACUCCI G., KLEIN M. L., MCDONALD I. R., *A Molecular Dynamics Study of the Lattice Vibrations of Sodium Chloride*, in "Journal de Physique Lettres", 36, 4, 1975, pp. 97-100.

KADANOFF L. et al., *Static Phenomena near Critical Points: Theory and Experiments*, in "Review of Modern Physics", 39, 1967, pp. 395-431.

KALOS M. H., *Monte Carlo Calculations of the Ground State of Three- and Four-Body Nuclei*, in "Physical Review", 128, 4, 1962, pp. 1791-5.

KALOS M. H., LEVESQUE D., VERLET L., *Helium at Zero Temperature with Hard-Sphere and other Forces*, in "Physical Review A", 9, 5, 1974, pp. 2178-95.

KAPRAL R., CICCOTTI G., *Molecular Dynamics: an Account of Its Evolution*, in C. Dykstra et al., *Theory and Applications of Computational Chemistry*, Elsevier, Amsterdam 2005, pp. 425-41.

KARPLUS M., *Carl Mathew Moser*, in Carl Moser Symposium, CECAM Symposium (Lyon, May 27, 2005), CECAM, Lyon 2005.

ID., *Carl Moser and CECAM*, in *Models for Protein Dynamics: 1976-2016*, CECAM Workshop (Lausanne, February 15-18, 2016), CECAM HQ-EPFL, Lausanne 2016, pp. 2-6.

KATAOKA Y., KLEIN M. L., *Shuichi Nosé*, in "Physics Today", 59, 2, 2006, pp. 67-8.

KECK J. C., *Variational Theory of Chemical Reaction Rates Applied to Three-Body Recombinations*, in "The Journal of Chemical Physics", 32, 4, 1960, pp. 1035-50.

KIRKWOOD J. C., MAUN E. K., ALDER B. J., *Radial Distribution Functions and the Equation of State of a Fluid Composed of Rigid Spherical Molecules*, in "The Journal of Chemical Physics", 18, 8, 1950, pp. 1040-7.

KOYRÉ A., *From the Closed World to the Infinite Universe*, Johns Hopkins Press, Baltimore (1957)

KOYRÉ A., *From the Closed World to the Infinite Universe*, Johns Hopkins Press, Baltimore 1957.

KUHN T, *The Copernican Revolution: Planetary Astronomy in the Development of Western Thought*, Harvard University Press, Cambridge, Mass. 1957.

ID., *The Structure of Scientific Revolutions*, Chicago University Press, Chicago 1962.

ID., *Dal mondo del pressappoco all'universo dellaprecisione*, Einaudi, Torino 1967. KUHN T, *The Copernican Revolution: Planetary Astronomy in the Development of Western Thought*, Harvard University Press, Cambridge (MA) 1957 (trad. it. *La rivoluzione copernicana*, Einaudi, Torino 1972).

ID., *The Structure of Scientific Revolution*, Chicago University Press, Chicago 1962 (trad. it. *La struttura delle rivoluzioni scientifiche*, Einaudi, Torino 1969).

LEBOWITZ J. L., PERCUS J. K., VERLET L., *Ensemble Dependence of Fluctuations with Application to Machine Computations*, in "Physical Review", 153, 1, 1967, pp. 250-4.

LEES A. W., EDWARDS S. F., *The Computer Study of Transport Processes under Extreme Conditions*, in "The Journal of Physics C", 5, 15, 1972, pp. 1921-8.

LEKKERKERKER H. N. W., *Daan Frenkel and the Spinoza Prize*, in "SIMU Newsletter", 2, 2000, p. 1 (https://www.researchgate.net/publication/267979729_SIMU_Challenges_in_Molecular_ Simulations_Bridging_the_Length-_and_Timescales_gap_Volume_2; ultima consultazione 22 aprile 2020).

LEVESQUE D., HANSEN J. P., *The Origin of Computational Statistical Mechanics in France*, in "The European Physics Journal H", 44, 1, 2019, pp. 37-46.

LEVITT M., WARSHEL A., *Computer Simulation of Protein Folding*, in "Nature" 253, 5494, 1975, pp. 694-8.

MAC KERNAN D., *Feature Interview, Michele Parrinello*, in "SIMU Newsletter", 2, 2000, p. 7 (https://www.researchgate.net/publication/267979729_SIMU_Challenges_in_Molecular_ Simulations_Bridging_the_Length_and_Times- cales_gap_Volume_2).

ID., *Interview, Kurt Binder*, in "SIMU Newsletter", 3, 2001, pp. 7-30 (https://www.academia.edu/ i5893770/SIMU_newsletter_Volume_3).

MAC KERNAN D., MARESCHAL M., *Interview with Berni Alder*, in "SIMU Newsletter", 4, 2002 (https://www.researchgate.net/publication/267979976_SIMU_Challenges_in_Molecular_ Simulations_Bridging_the_Length_and_Timescales_gap_Volume_4).

MACUGLIA D., ROUX B., CICCOTTI G., *Sense Experiences and "Necessary Simulations": Four Centuries of Scientific Change from Galileo to Fundamental Computer Simulations*, in "KNOW: A Journal on the Formation of Knowledge", 4:1, 2020, pp. 63-87.

MANSIGH M. A., *The Early Years of Molecular Dynamics and Computers at UCRL, LRL, LLL, and LLNL*, in E. Schwegler, B. M. Rubenstein, S. B. Libby (eds.), *Advances in the Computational Sciences: Proceedings of the Symposium in Honor of Dr Berni Alder's 90th Birthday*, Proceedings of the Symposium (Livermore, August 20, 1985), World Scientific, Singapore 2015, pp. 176-83.

MARESCHAL M., *Early Years of Computational Statistical Mechanics*, in "European Physical Journal H", 43, 293, 2018, pp. 293-302.

ID., *From Varenna (1971) to Como (1995): Kurt Binder's Long Walk in the Land of Criticality*, in "European Physics Journal H", 44, 2019, pp. 161-79.

MCDONALD I. R., *NpT-ensemble Monte Carlo Calculationsfor Binary Liquid Mixtures*, in "Molecular Physics", 23, 1, 1972, pp. 41-58.

MCDONALD I. R., SINGER K., *Calculation of Thermodynamic Properties of Liquid Argon from Lennard-Jones Parameters by a Monte Carlo Method*, in "Discussions of Faraday Society", 43, 1967, pp. 40-9.

ID., *Machine Calculation of Thermodynamic Properties of a Simple Fluid at Supercritical Temperatures*, in "The Journal of Chemical Physics", 47, II, 1967, pp. 4766-72.

ID., *Examination of the Adequacy of the 12-6 Potential for Liquid Argon by Means of Monte Carlo Calculations*, in "The Journal of Chemical Physics", 50, 6, 1969, pp. 2308-15.

MCMILLAN W. L., *Ground State of Liquid He⁴*, in "Physical Review A", 138, 2, 1965, pp. 442-51.

METROPOLIS N., *The Beginning of the Monte Carlo Method*, in "Los Alamos Science", Special Issue, 15, 1987, pp. 125-30.

METROPOLIS N., HOWLETT J., ROTA G. C. (eds.), *A History of Computing in the Twentieth Century*, Academic Press, New York 1980.

METROPOLIS N., ULAM S., *The Monte Carlo Method*, in "Journal of the American Statistical Association", 44, 247, 1949, pp. 335-41.

METROPOLIS N. *et al.*, *Equation of State Calculations by Fast Computing Machine*, in "The Journal of Chemical Physics", 21, 6, 1953, pp. 1087-92.

MICHAEL G., *An Interview with Bernie Alder*, March 5, 1997 (http://www.com-puter-history.info/PageI.dir/pages/Alder.html).

MULLER-KRUMBHAAR H., BINDER K., *Dynamic Properties of the Monte Carlo Method in Statistical Mechanics*, in "Journal of Statistical Physics", 8, I 1973, pp. 1-24.

NOSÉ S., *A Molecular-Dynamics Method for Simulations in the Canonical Ensemble*, in "Molecular Physics", 52, 2, 1984, pp. 255-68.

ID., *A Unified Formulation of the Constant Temperature Molecular-Dynamics Methods*, in "The Journal of Chemical Physics", 81, 1, 1984, pp. 511-9.

ONSAGER L., *Crystal Statistics. I. A Two-Dimensional Model with an Order-Disorder Transition*, in "Physical Review", 65, 3-4, 1944, pp. 117-49.

ORLANDINI S., MELONI S., CICCOTTI G., *Hydrodynamics from Dynamical Nonequilibrium MD*, in "AIP Conference Proceedings", 1332, 77, 2011.

PACI E., *Foreword*, in *Models for Protein Dynamics: 1976-2016*, CECAM Workshop (Lausanne, February 15-18, 2016), CECAM HQ-EPFL, Lausanne 2016.

PANAGIOTOPOULOS A. Z., *Direct Determination of Phase Coexistence Properties of Fluids by Monte Carlo Simulation in a New Ensemble*, in "Molecular Physics", 61, 4, 1987, pp. 813-26.

ID., *Adsorption and Capillary Condensation of Fluids in Cylindrical Pores by Monte Carlo Simulation in the Gibbs Ensemble*, in "Molecular Physics", 62, 3, 1987, pp. 701-19.

PANAGIOTOPOULOS A. Z., SUTER U. W., REID R. C., *Phase Diagrams of Nonideal Fluid Mixtures from Monte Carlo Simulation*, in "Industrial & Engineering Chemistry Fundamentals", 25, 4, 1986, pp. 525-35.

PANAGIOTOPOULOS A. Z. *et al.*, *Phase Equilibria by Simulation in the Gibbs Ensemble: Alternative Derivation, Generalization and Application to Mixture and Membrane Equilibria*, in "Molecular Physics", 63, 4, 1988, pp. 527-45.

PARRINELLO M., RAHMAN A., *Crystal Structure and Pair Potentials: A Molecular-Dynamics Study*, in "Physical Review Letters", 45, 14, 1980, pp. H96-9.

PERCUS J. K. (ed.), *The Many-Body Problem*, Proceedings of the Symposium at Stevens Institute of Technology (Hoboken, NJ, January 28-29, 1957), Interscience Publishers, New York-London 1963.

POINCARÉ H, *La valeur de la science*, Flammarion, Paris 1890.

POLLOCK E. L., CEPERLEY D. M., *Simulation of Quantum Many-Body Systems by Path Integral Methods*, in "Physical Review B", 30, 5, 1984, pp. 2555-68.

RAHMAN A, *Correlation in the Motion of Atoms in Liquid Argon*, in "Physical Review A", 2, 136, 1964, pp. 405-11.

RAHMAN A., STILLINGER F., *Molecular Dynamics Study of Liquid Water*, in "The Journal of Chemical Physics", 55, 7, 1971, pp. 3336-59.

REIF F., *Statistical Physics: Berkeley Physics Course*, vol. 5, McGraw-Hill, New York 1965.

RICE S. A., STILLINGER F. H., *John Gamble Kirkwood (1907-1959): A Biographical Memoir*, in *Biographical Memoirs*, vol. 77, The National Academy Press, Washington DC 1999.

ROSENBLUTH M. N., ROSENBLUTH A. W., *Further Results on Monte Carlo Equations of State*, in "The Journal of Chemical Physics", 22, 5, 1954, pp. 881-4.

ID., *Monte Carlo Calculation of the Average Extension of Molecular Chains*, in "The Journal of Chemical Physics", 23, 2, 1955, pp. 356-9.

RYCKAERT J. P., CICCOTTI G., BERENDSEN H. J. C., *Numerical Integration of the Cartesian Equation of Motion of a System with Constraints: Molecular Dynamics of N-alkanes*, in "Journal of Computational Physics", 23, 3, 1977, pp. 327-41.

SARIBAN A., BINDER K., *Critical Properties of the Flory-Huggins Lattice Model of Polymer Mixtures*, in "The Journal of Chemical Physics", 86, 10, 1987, pp.585-973.

SCHEYER M., *Asylum*, Profile Books, London 2016.

SCHWEGLER E., RUBENSTEIN B. M., LIBBY S. B. (eds.), *Advances in the Computational Sciences: Symposium in Honor of Dr Berni Alder's 90th Birthday* (Lawrence Livermore National Laboratory, August 20, 2015), World Scientific, Singapore 2017.

SEGRÉ E., *From X-ray to Quarks*, W. H. Freeman, San Francisco 1980.

SEN K., SASTRY S., *Aneesur Rahman: A Pioneer in Computational Physics*, in "Resonance", 19, 2014, pp. 671-83.

SINGER K., TAYLOR A., SINGER J. V. L., *Thermodynamic and Structural Properties of Liquids Modelled by '2-Lennard-Jones Centres' Pair Potentials*, in "Molecular Physics", 33, 6, 1977, pp. 1757-95.

SINGER K., WOODCOCK I. V., *Thermodynamic and Structural Properties of Liquid Ionic Salts obtained by Monte Carlo Computation. 1. Potassium Chloride*, in "Transactions of the Faraday Society", 67, 1971, pp. 12-30.

SMITH E. B., LEA K. R., *A Monte Carlo Equation of State for Mixtures of Hard-Sphere Molecules*, in "Nature", 186, 4726, 1960, p. 714.

SPRIK M. *et al.*, *Tribute to Michael L. Klein: Scientist, Teacher, and Mentor*, in "Journal of Physical Chemistry B", 110, 8, 2006, pp. 3451-3.

STANLEY H. E., *Introduction to Phase Transitions and Critical Phenomena*, Clarendon Press, Oxford 1971.

THIELE E., *Equation of State for Hard Spheres*, in "The Journal of Chemical Physics", 39, 2, 1963, pp. 474-9

TORRIE G., VALLEAU J. P., *Monte Carlo Free Energy Estimates Using Non-Boltz-mann Sampling: Application to the Sub-Critical Lennard Jones Fluid*, in "Chemical Physics Letters", 28, 4, 1974, pp. 578-81.

ID., *Nonphysical Sampling Distributions in Monte Carlo Free-Energy Estimation: Umbrella Sampling*, in "Journal of Computational Physics", 23, 2, 1977, pp. 187-99.

TORRIE G., VALLEAU J. P., BAIN A., *Monte Carlo Estimation of Communal Entropy*, in "The Journal of Chemical Physics", 58, 12, 1973, pp. 5479-83.

VALLEAU J. P., CARD D. N., *Monte Carlo Estimation of the Free Energy by Multistage Sampling*, in "The Journal of Chemical Physics", 57, 12, 1972, pp. 5457-62.

VAN GUNSTEREN W. F., *The Roots of Biomolecular Simulation*, in *Models for Protein Dynamics: 1976-2016*, CECAM Workshop (Lausanne, February 15-18, 2016), CECAM HQ-EPFL, Lausanne 2016, pp. 7-10.

VERLET L., *Computer "Experiments" on Classical Fluids. I. Thermodynamical Properties of Lennard-Jones Molecules*, in "Physical Review", 159, 1, 1967, pp. 98-103.

ID., *The Origins of Molecular Dynamics*, in J. P. Hansen, G. Ciccotti, H. J. C. Berendsen (eds.), *In Memoriam: Aneesur Rahman 1927-1987*, CECAM Meeting (Orsay, September II, 1987), CECAM, Orsay 1987, pp. 6-8.

ID., *Chimères et Paradoxes*, Les Éditions du Cerf, Paris 2007.

WERTHEIM M, *Exact Solution of the Percus-Yevick Integral Equation for Hard Spheres*, in "Physical Review Letters", 10, 8, 1963, pp. 321-3.

WILSON K. G., *Renormalization Group and Critical Phenomena. I. Renormalization Group and the Kadanoff Scaling Picture*, in "Physical Review B", 4, 9, 1971, pp. 3174-83

WOOD W. W., *Early History of Computer Simulations in Statistical Mechanics*, in G. Ciccotti, W. G. Hoover (eds.), *Molecular-Dynamics Simulation of Statistical-Mechanical Systems*, Proceedings of the "Enrico Fermi" International Summer School of Physics, Course XCVII (Varenna, July 23-August 2, 1985), North-Holland, Amsterdam 1986, pp. 3-14.

WOOD W. W., JACOBSON J. D., *Preliminary Results from a Recalculation of the Monte Carlo Equation of State of Hard Spheres*, in "The Journal of Chemical Physics", 27, 5, 1957, pp. 1207-8.

YAMAMOTO T., *Quantum Statistical Mechanical Theory of the Rate of Exchange Chemical Reactions in the Gas Phase*, in "The Journal of Chemical Physics", 33, 1, 1960, pp. 281-9.

# Index

© Springer Nature Switzerland AG 2020
G. Battimelli et al., *Computer Meets Theoretical Physics*, The Frontiers Collection,
https://doi.org/10.1007/978-3-030-39399-1

Printed in the United States
by Baker & Taylor Publisher Services